Palgrave Studies in European Union Politics

Edited by: **Michelle Egan**, American University USA, **Neill Nugent**, Manchester Metropolitan University, UK and **William Paterson OBE**, University of Aston, UK.

Editorial Board: **Christopher Hill**, Cambridge, UK, **Simon Hix**, London School of Economics, UK, **Mark Pollack**, Temple University, USA, **Kalypso Nicolaïdis**, Oxford UK, **Morten Egeberg**, University of Oslo, Norway, **Amy Verdun**, University of Victoria, Canada, **Claudio M. Radaelli**, University of Exeter, UK, **Frank Schimmelfennig**, Swiss Federal Institute of Technology, Switzerland

Following on the sustained success of the acclaimed *European Union Series*, which essentially publishes research-based textbooks, *Palgrave Studies in European Union Politics* publishes cutting edge research-driven monographs.

The remit of the series is broadly defined, both in terms of subject and academic discipline. All topics of significance concerning the nature and operation of the European Union potentially fall within the scope of the series. The series is multidisciplinary to reflect the growing importance of the EU as a political, economic and social phenomenon.

Titles include:

Ian Bache and Andrew Jordan *(editors)*
THE EUROPEANIZATION OF BRITISH POLITICS

Thierry Balzacq *(editor)*
THE EXTERNAL DIMENSION OF EU JUSTICE AND HOME AFFAIRS
Governance, Neighbours, Security

Falk Daviter
POLICY FRAMING IN THE EUROPEAN UNION

Renaud Dehousse *(editor)*
THE 'COMMUNITY METHOD'
Obstinate or Obsolete?

Kenneth Dyson and Angelos Sepos *(editors)*
WHICH EUROPE?
The Politics of Differentiated Integration

Michelle Egan, Neill Nugent, and William E. Paterson *(editors)*
RESEARCH AGENDAS IN EU STUDIES
Stalking the Elephant

Kevin Featherstone and Dimitris Papadimitriou
THE LIMITS OF EUROPEANIZATION
Reform Capacity and Policy Conflict in Greece

Stefan Gänzle and Allen G. Sens *(editors)*
THE CHANGING POLITICS OF EUROPEAN SECURITY
Europe Alone?

Eva Gross
THE EUROPEANIZATION OF NATIONAL FOREIGN POLICY
Continuity and Change in European Crisis Management

Adrienne Héritier and Martin Rhodes *(editors)*
NEW MODES OF GOVERNANCE IN EUROPE
Governing in the Shadow of Hierarchy

Wolfram Kaiser, Brigitte Leucht, and Michael Gehler
TRANSNATIONAL NETWORKS IN REGIONAL INTEGRATION
Governing Europe 1945–83

Hussein Kassim and Handley Stevens
AIR TRANSPORT AND THE EUROPEAN UNION
Europeanization and its Limits

Robert Kissack
PURSUING EFFECTIVE MULTILATERALISM
The European Union, International Organizations and the Politics of Decision Making

Katie Verlin Laatikainen and Karen E. Smith (*editors*)
THE EUROPEAN UNION AND THE UNITED NATIONS
Intersecting Multilateralisms

Esra LaGro and Knud Erik Jørgensen (*editors*)
TURKEY AND THE EUROPEAN UNION
Prospects for a Difficult Encounter

Ingo Linsenmann, Christoph O. Meyer and Wolfgang T. Wessels (*editors*)
ECONOMIC GOVERNMENT OF THE EU
A Balance Sheet of New Modes of Policy Coordination

Hartmut Mayer and Henri Vogt (*editors*)
A RESPONSIBLE EUROPE?
Ethical Foundations of EU External Affairs

Philomena Murray (*editor*)
EUROPE AND ASIA
Regions in Flux

Daniel Naurin and Helen Wallace (*editors*)
UNVEILING THE COUNCIL OF THE EUROPEAN UNION
Games Governments Play in Brussels

David Phinnemore and Alex Warleigh-Lack
REFLECTIONS ON EUROPEAN INTEGRATION
50 Years of the Treaty of Rome

Sebastiaan Princen
AGENDA-SETTING IN THE EUROPEAN UNION

Carolyn Rowe
REGIONAL REPRESENTATION IN THE EU
Between Diplomacy and Interest Mediation

Emmanuelle Schon-Quinlivan
REFORMING THE EUROPEAN COMMISSION

Roger Scully and Richard Wyn Jones (*editors*)
EUROPE, REGIONS AND EUROPEAN REGIONALISM

Asle Toje
AFTER THE POST-COLD WAR
The European Union as a Small Power

Richard G. Whitman and Stefan Wolff (*editors*)
THE EUROPEAN NEIGHBOURHOOD POLICY IN PERSPECTIVE
Context, Implementation and Impact

Richard G. Whitman (*editor*)
NORMATIVE POWER EUROPE
Empirical and Theoretical Perspectives

Palgrave Studies in European Union Politics
Series Standing Order ISBN 978–1–4039–9511–7 (hardback) and
ISBN 978–1–4039–9512–4 (paperback)

You can receive future titles in this series as they are published by placing a standing order.
Please contact your bookseller or, in case of difficulty, write to us at the address below with
your name and address, the title of the series and the ISBN quoted above.

Customer Services Department, Macmillan Distribution Ltd, Houndmills, Basingstoke,
Hampshire RG21 6XS, England

Policy Framing in the European Union

Falk Daviter

Junior Professor of Public Policy and Administration, Department of Economics and Social Sciences, University of Potsdam, Germany

First published 2011 by
PALGRAVE MACMILLAN

Palgrave Macmillan in the UK is an imprint of Macmillan Publishers Limited,
registered in England, company number 785998, of Houndmills, Basingstoke,
Hampshire RG21 6XS.

Palgrave Macmillan in the US is a division of St Martin's Press LLC,
175 Fifth Avenue, New York, NY 10010.

Palgrave Macmillan is the global academic imprint of the above companies
and has companies and representatives throughout the world.

Palgrave® and Macmillan® are registered trademarks in the United States,
the United Kingdom, Europe and other countries

ISBN 978-0-230-27778-6   hardback

This book is printed on paper suitable for recycling and made from fully
managed and sustained forest sources. Logging, pulping and manufacturing
processes are expected to conform to the environmental regulations of the
country of origin.

A catalogue record for this book is available from the British Library.

Library of Congress Cataloging-in-Publication Data

Daviter, Falk, 1975–
  Policy framing in the European Union / Falk Daviter.
    p. cm.
  Includes index.
  ISBN 978-0-230-27778-6 (hardback)
  1. Policy sciences. 2. Biotechnology–Government policy–European Union
  countries. I. Title.

H97.D395 2011
660.6094–dc23   2011021097

10   9   8   7   6   5   4   3   2   1
20 19 18 17 16 15 14 13 12 11

Printed and bound in the United States of America

Contents

Acknowledgements

The research for this book was completed at the European University Institute (EUI) in Florence. During this time, I was very fortunate to share the splendid seclusion of the Badia Fiesolana with an exceptionally inspiring group of people and to find a home at Via Faentina 13. Out of the many who contributed to this research project, I wish to thank some in particular. Adrienne Héritier provided professional guidance, personal encouragement and, above all, an unwavering sense of direction. I am deeply grateful for the interest she took in my work. Her support was an irreplaceable part of the process of writing this book. Ellen Immergut, Claudio Radaelli and Stefano Bartolini provided highly valued and detailed criticism of earlier versions of my manuscript. I greatly appreciate their time and insight. Their comments helped me to turn my research into a book. Much of this research would not have been possible without the assistance of the EUI library staff. Their professionalism and commitment were invaluable. During several trips to Brussels, members of the European Commission as well as other EU policy experts generously shared their time and perspective, and allowed me to access a wealth of data.

I gratefully acknowledge permission to use material from the previously published article 'Policy Framing in the European Union', *Journal of European Public Policy* 14(4), June 2007, pp. 654–666.

List of Abbreviations

BCC	Biotechnology Coordination Committee
BEUC	European Consumers' Organisation
BIODOC	Biotechnology Documentation Centre
BRIC	Biotechnology Regulations Inter-Service Committee
BSC	Biotechnology Steering Committee
BSE	Bovine spongiform encephalopathy
CEFIC	Council of European Chemical Industry
CJD	Creutzfeldt-Jakob Disease
CUBE	Concertation Unit for Biotechnology in Europe
DG	Directorate-General
DNA	Deoxyribonucleic acid
EBCG	European Biotechnology Co-ordination Group
EC	European Commission
ECJ	European Court of Justice
ECRAB	European Committee on Regulatory Aspects of Biotechnology
EFPIA	European Federation of Pharmaceutical Industry Associations
EFSA	European Food Safety Authority
EIS	European Information Service
EPP	European People's Party
ESNBA	European Secretariat of National Bioindustry Associations
EU	European Union
FEBC	Forum for European Bioindustry Coordination
GMM	Genetically modified microorganism
GMO	Genetically modified organism
MEP	Member of the European Parliament
NGO	Non-governmental organisation
OECD	Organisation for Economic Co-operation and Development
PES	Party of European Socialists
rDNA	recombinant deoxyribonucleic acid

SAGB   Senior Advisory Group on Biotechnology
SANCO   Directorate General for Health and Consumer Affairs
SEA   Single European Act

1
Introduction

The main point of this book is that issues drive the policy process. Traditionally, research in political science views policy choices as resulting from factors that are distinct from the issues under consideration. This book turns the argument around. Policy issues are inherently complex and ambiguous. How to evaluate policy alternatives often remains contentious. The framing of policy issues affects the processing of political interests and ideas and their expression in policy choices. What policy makers perceive to be at stake in a policy issue at a particular time affects the political alignment of actors and the conflict and consolidation of interests in the policy process. Political actors seek to control the flow and structure of policy issues, because issues are the currency of politics. At a time when framing arguments are attracting increasing attention in policy research, this book provides the first comprehensive study of policy framing in the EU. It introduces the conceptual and theoretical tenets of framing analysis and shows how this analytical lens can offer a unique perspective on current issues in the study of EU legislative politics and policy making.

Policy framing and the study of politics

Students of policy making have long argued that political issues are not external to the process of political decision making – they are not 'out there'. Every policy issue can be defined in a number of different ways. Conflicting perceptions of what policy issues are about are often difficult or impossible to reconcile. Yet the perception of

1

an issue which prevails, the facet or dimension of the issue which dominates policy debates at a given time, can substantially influence political choices. Only recently have political scientists turned to focus more closely on the question of how the definition of political issues affects the processing of political ideas and political demands in policy making. This literature refers to 'framing' as the process of selecting and emphasising aspects of an issue according to an overriding evaluative or analytical criterion. Policy frames identify what is at stake in an issue. Students of policy framing ask how the framing of choices influences the way the issue is processed by the political system. How does framing affect which actors and institutions play a role during policy drafting and deliberation? How does the framing of the issues influence which interests find expression in policy choices?

Initially, the study of political issue definition was treated as an integral part of the more established study of agenda setting in political science and policy analysis (Cobb and Elder 1971; Rochefort and Cobb 1993, 1994; Cobb and Ross 1997). Following classic studies such as Nelson's (1984) work on the issue of child abuse, however, the political representation of policy issues received increasing attention (e.g. Stone 1989; Petracca 1992; Peters 2005). The effects of issue definitions were soon found to go beyond the initiation of the policy process and to serve several functions in the policy making process. Dery's (2000) study of a faltering social protest movement exemplifies research that highlights more ambivalent aspects of the relationship between agenda setting and issue definition. He shows how the failure to control problem representations after the initial agenda setting success can render the advocacy for a specific cause ineffective. 'To legitimize an issue', Dery (2000: 37) concludes, 'is not the same as to legitimize demands.' Other authors have taken steps towards conceptualising problem definitions as the outcome of policy making and as an integral part of adopted policy. From this perspective, issue frames have been found to influence policy dynamics over the long run. Research in this tradition emphasises how the specific representation and delineation of policy issues shapes the formation and organisation of interests, at times restructuring political constituencies in the process (e.g. Schmitter 1992: 379; Pierson 1993).

Most research attention, however, is devoted to analysing policy frames as a 'weapon of advocacy and consensus' (Weiss 1989: 117) during the political decision making process itself. Research from this perspective typically breaks with the traditional notion that the definition of policy issues can be properly understood as the initial phase of a structured policy process or cycle. Here the definition of policy issues is seen as lying 'at the heart of the action itself' (Weiss 1989: 98) rather than being ascribed to the pre-decisional realm of politics. This type of framing research owes much to Schattschneider's (1957, 1960) conception of politics. From this perspective, the definition of policy choices in the political process is intimately linked to the emergence of political conflict and the evolving structure of political competition (see also Mair 1997: 949). The definition of policy issues and alternatives structure ensuing political conflicts because they fix the attention of the public, influence the formation and organisation of interests and shape political coalitions and alliances. Who has a say in the political decision making process, Schattschneider (1960: 102) argues, depends on 'what the game is about.' More recently, Riker (1986: 150–151) identified the manipulation of issue definitions as one of the most frequently employed strategies in political discourse and claimed that 'most of the great shifts in political life' are caused by the reframing of the issues at stake. As part of his work on heresthetics, Riker advances framing arguments by elaborating how the manipulation of issue definitions can reshuffle majorities. When new dimensions of the issue are highlighted during political debates, Riker (1986: 145) argues, voters' assessments of the alternatives restructure in such a way that a stable equilibrium is easily lost and new majority-minority divisions emerge. Yet while the causal effects of issue framing on individual decision making have been studied extensively, the ways in which framing dynamics play out in complex policy making environments is less well understood. As Baumgartner and Jones (1991, 2002: 298) point out, frame manipulation rarely goes uncontested. In democratic politics, disadvantaged political actors often challenge dominant issue definitions by raising or emphasising new dimensions of the choice. As a result, the study of stable and systematic effects of policy framing on political choices continues to present theoretical challenges. Current research on policy framing predominantly follows the work of Baumgartner and Jones (1991, 2002; see also Jones 1994b). Their research emphasises that one key to understanding the dynamics of policy framing and reframing lies in analysing how framing effects interact with the institutional organisation of politics. According to this perspective, the institutional channels, or policy venues, through which political issues are processed, focus the decision maker's attention on a simplified image of complex policy choices and thereby exert bias towards the inclusion of certain types of information and interests over others (Baumgartner and Jones 1991). The nature of framing effects will thus differ from one policy making system to the next. How policy framing plays out in the complex and fragmented system of EU policy making is the central question addressed in this book. In addressing these questions, this book focuses on policy issues as the central units of analysis. In contrast to policy studies that place emphasis on substantive questions of problem solving, however, the arguments developed in this book are about the politics of policy making. How policy issues affect the scope and nature of political conflict and competition in the EU is a central concern of this book. This conceptualisation of the EU policy process yields insights that are often at odds with standard accounts of EU politics, and it sheds new light on the analysis of the EU as a political system. At the same time, the empirical and theoretical findings offer a unique perspective within the larger theoretical context of the study of framing effects in political research. This book thus provides fresh insights into the dynamics of policy framing more generally and its specific effects on supranational decision making in the EU.

Why EU studies need a framing perspective

The following section provides a more detailed argument of why studies of EU policy making can gain analytical leverage by placing more emphasis on policy issues as a central unit of analysis and by theorising framing effects on supranational policy dynamics more comprehensively. While a systematic analysis of the policy making system of the European Union from a framing perspective has so far not been advanced, the centrality of policy issues and political problem definitions during EU agenda setting and policy formulation has been highlighted repeatedly (e.g. Peters 1994, 2001; Princen 2007, 2009). Due to its vertical and horizontal fragmentation, the EU offers an unusually high number of access points for agenda setters (e.g. Princen and Kerremans 2008). As a result, issues are regularly taken up and processed simultaneously, but independently, at various levels of EU decision making. Party coercion or political programmes that could provide consistency across the institutions of the EU are often lacking. Instead, the following discussion will show that the EU policy making system is characterised by contested competencies and competing constituencies. Its policy process is riddled with conflicting organisational logics and incoherently reconciled representational functions. Research on EU legislative policy making from diverse theoretical perspectives has essentially converged around the conclusion that it is 'frequently difficult to predict how key actors will align themselves on any given issue or which battle along which cleavage will matter most in determining outcomes' (Peterson 2001: 292–293). Because issues in EU politics rarely enter the political agenda neatly wrapped, as Peters (1994, 2001) argues, political dynamics are largely endogenous to the processing of policy issues at the supranational level. Packaging policy initiatives is therefore a highly politicised process. While recent advances in EU studies from diverse analytical perspectives increasingly underscore the relevance of this finding, students of the EU policy process still grapple with the theoretical repercussions. The following overview begins by highlighting problems with studies that presuppose stable and exogenous patterns of political conflict in EU policy making and then successively broadens the scope of analysis to incorporate insights gained from the study of the organisational level of EU politics and policy making.

The search for the EU political space

In recent years, the notion of the EU political space (or conflict space), a concept that encompasses *ex ante* assumptions about the dimensionality of EU policy choices, has attracted increasing attention. Many conclusions drawn from formal models of EU legislative politics are in fact derived directly from these assumptions. As a result, one of the proponents of this literature insists that until recently available findings were 'still conjectures, which might not be empirically true' (Steunenberg 2000: 369; see also Selck 2004a for a critical review from this viewpoint). Different conceptualisations of the EU conflict space indeed give rise to widely differing conclusions about likely political outcomes (see also Marks and Steenbergen 2002). As empirical research designed to inform the specification of spatial models of the EU is becoming more common, many of the core assumptions of the dominant models have been called into question. As a result, arguments concerning reasonably invariant patterns of EU politicisation have started to unravel.

In most of the spatial literature on the EU, the 'supranational scenario' of the EU political space, an assumption according to which actors align along a single dimension depending on their preference for more or less EU integration, has been the dominant approach and remains the central point of reference (e.g. Steunenberg 1994; Crombez 1996, 1997; see also Tsebelis and Garrett 2000: 15–26 for a summary of the argument). In a recent overview, even sceptics of this one-dimensional model like Tsebelis and Garrett (2000: 26) consider it 'a reasonable characterisation of much EU legislative politics in the last decade.' Despite its appealing simplicity, however, studies of actual EU decision making have found little support for this scenario. Recent research based on the analysis of 70 policy proposals concludes instead that the 'inspection of actor alignments does not support the supranational scenario' (Thomson et al. 2004: 252). These findings are further supported by Kreppel's (2002) empirical analysis of Parliament amendments, which were rejected by the Commission. The level of policy disagreement documented by this research calls a central feature of the supranational scenario in question, namely the general prediction of an alliance between the two supranational EU institutions on most policy issues based on their common preference for policy choices that further EU integration. A more general empirical refutation of one-dimensional spatial models of EU politics can be found in Selck (2004b). Empirical studies also find little support for the main alternative scenario, a model originally proposed by Tsebelis (1994) and reiterated in a more recent version of the original argument (Tsebelis and Garrett 2001). According to this second scenario, the authors explain, 'many important policy disputes in the contemporary EU appear to take place in a two-dimensional policy space – one dimension describes actors' preferences for more or less regional integration, and the other is more akin to the traditional left–right cleavage' (2001: 377; see also Hix et al. (2006) on the left–right dimension in the European Parliament). Based on this perception of the political space, Tsebelis and Garrett further hypothesise that the Council and Parliament will take the two extreme positions on the integration dimension, with the Commission as a less extremely positioned ally of the Parliament in favour of more integration. Along the left–right dimension, the authors assume that the Commission will largely reflect the views of the current Council majority. Yet empirical research of actor alignments inside the Council reveals that the left–right dimension is weak and regularly overridden or reconfigured by more specific interest constellations and conflicts (see Zimmer et al. 2005: 411). Thomson et al. (2004: 254) equally find voting in the Council to correlate with the left–right positions of the respective member states at a level so low that the results could easily be random. Other research teams confirm that classifying 'member states according to their overall preference patterns fails due to the inconsistencies of their preference profiles' (Zimmer et al. 2005: 418). The predominant impression, Zimmer and her collaborators summarise, is that 'no unifying political space within the European Union exists' (2005: 405). As the underlying dimensions of EU policy choices have drawn increasing attention, Pollack (1997a: 131; see also Aspinwall and Schneider 2000: 23) cautions that *post hoc* manipulation of the models to account for ever newer scenarios of the EU political space can turn the declared goal of rigorous theory-testing into mere exercises in 'curve fitting'.

Far from finding support for many of the scenarios advanced by spatial theorists, researchers find shifting patterns of coalitions as traditionally described in more empirically informed studies of EU politics (e.g. Nugent 1999: 474). Main findings of the empirical investigation of EU spatial models thus points to the 'weakness of the dimensionality of the EU political space' (Thomson et al. 2004: 257). Few cases of EU political decision making appear to reflect exogenous actor preferences in a stable and low-dimensional political space. Instead, political preferences are regularly found to form or restructure at the EU level. In order to account for such dynamics, Ringe (2005) proposes to start out from a two-dimensional model of the EU political space, but argues that the actual policy choices must be traced back to focal points that establish which specific aspect of a legislative proposal dominates the perceptions of the political actors. A similar extension of the standard spatial argument is also discussed by Rittberger (2000), who shows how the Parliament can raise previously latent issue dimensions and thereby recast political conflicts. In short, preconceived notions of an EU conflict space have proven to be a somewhat perilous point of departure for theories of EU politics and policy making. If the EU conflict space is weak, theories of EU politics must instead turn to study the process of its formation. Students of EU legislative politics have thus proposed several ways of broadening the scope of the inter- and intra-institutional analysis. As Ringe (2005: 743) concludes, the unitary actor assumption of spatial models appears insufficient to account for preference formation inside Council, Parliament and Commission, a point Hörl et al. (2005) elaborate in greater detail. Such a research agenda would reassess the political construction of interests at the organisational level of European politics. Another way to start, as Zimmer et al. (2005: 418) point out, would be by gaining 'an intuitive grasp of the role of the European Commission in the European conflict space.' The promises and pitfalls of both research strategies are addressed in turn.

The contested centre

With respect to the process of policy formation, the exclusive right of the European Commission to initiate legislative policy is one of the most striking features of the EU political system. Yet while the right of initiative gives the Commission a substantial legislative function under most articles of the treaty (see also Cini 2006), the Commission is strictly speaking not part of the legislature since it does not vote on legislative policy. Formal institutional analysis alone thus conveys a very inconclusive impression of the role of this political actor. Even within the mainstream of EU political analysis, evaluations of the role of the Commission in supranational policy making differ widely. Some literatures dismiss the right of initiative as insignificant and ascribe the Commission only marginal legislative influence. Other literatures conclude that the formal right of initiative is but a shadow of the actual power this player wields in EU political practice. Partially underlying the puzzling differences in how much importance is attributed to the role of the European Commission in EU policy formation are more general differences over the characterisation of the EU political order. While the Commission assumes important executive functions, it clearly defies a straightforward classification as the EU's executive. Instead, the Commission shares its executive decision making power with the Council and Parliament in a complex system of comitology (e.g. Pollack 2005: 75–154; Tallberg 2000; Dogan 1997) and depends almost entirely on the member states for implementation and enforcement. The Commission's legislative function, on the other hand, goes far beyond the phase of policy initiation guaranteed by the right of initiative. In the vast majority of issue areas the Commission 'enjoys a strong form of proposal power known as gatekeeping power, i.e. a monopoly on legislative initiative and the right to withdraw legislative proposals at any time' (Jupille 2007: 303). The Commission thus plays a role at various stages of the legislative process, and it can choose whether to accept or reject amendments made at previous decision points by the Council or Parliament, or whether to introduce new language.

Reviewing the bulk of research on EU agenda setting and policy formulation, Marks et al. (1996: 361) conclude: 'The European Commission is a critical actor in the policy initiation phase, whether one looks at formal rules or practice.' Conceptualisation of the ways in which formal rules and practice can account for EU policy choices however, remains tentative. Nugent (1997: 21) emphasises that the ways in which the Commission plays out its role only become apparent through research which covers 'the whole decision-making process, and in particular is directed towards the preparation of the decision-taking ground'. He insists that the 'nature of EU decision-making is so extremely complex... that the Commission has many opportunities to play roles and to exercise influence over and above its formal responsibilities'. As Nugent (1997: 10–11) notes, however, 'good strategic positioning is not enough'. Short of providing a fully explanatory approach, he specifies numerous resources and constraints, which help to define the possibilities for independent Commission agency. 'However, precisely how is has done this', Nugent (1995: 621) concludes, 'and, more particularly, how it has related its goals, its strategies and its tactics to its capacities is the subject matter for further work'. While Peterson (1995: 78, 86) equally subscribes to the view that in the EU, 'policy outcomes are shaped in crucial ways early in the policy-making process', he also notes the extent to which the previous predominance of EU theories originating in the field of international relations has left the link between policy formulation and policy outcomes in the EU largely unexplored.

In contrast, theoretical models that perceive of the Commission as a passive agent that depends on the member states (and, increasingly, Parliament) are among the most fully developed in studies of EU politics. Representing this perspective, Pollack conceptualises the Commission more as the target of intentional political agency rather than as a central political player in its own right (1996, 1997a, 1997b, 1998, 2001: 229–230). He acknowledges that Commission expertise, brokerage skills, institutional persistence and formal powers make it appear a potential candidate for successful political entrepreneur-ship (1997a: 126–127). But his research explores this argument only within limits. By assigning member states fixed and stable policy preferences, he describes the political autonomy of the Commission as a function of national interests, the distribution of information and the decision rules (e.g. 1997b: 124–128, 1998: 247–249, 2000: 8). Under favourable conditions, which are limited to cases characterised by high levels of uncertainty or 'diffuse interests' (1997b), Pollack (1997a: 130) concludes that the Commission 'may provide focal points around which the uncertain preferences of member governments can converge.' Yet in Pollack's interpretation, instances in which the Commission can successfully establish the political agenda are the exception to the rule.

This theoretical perspective leaves the Commission with little room to manoeuvre or manipulate the process of EU policy formation. It also invites criticism because his findings predominantly reflect the ontological and behavioural premises on which this research is based. Pollack (e.g. 1997a: 99) explores the question of whether the Commis-sion matters within a largely intergovernmental analytical framework that brackets the analysis of most areas of possible Commission influences in the supranational policy process. In theoretical terms, Pollack furthermore roots his analysis in formalistic conceptions of political agenda setting power that have come under intense scrutiny even among rational choice proponents (e.g. Moe 2005: 226–228). As a result, Pollack's work is bound to portray Commission agency as comparatively constrained and he points out that other approaches capture the full extent of relevant political dynamics far better (e.g. 1997a: 125). Schmidt (2000: 38, 56) similarly notes that the emphasis on the formal role of the Commission as the agenda setter fails to pro-vide conclusive insights, and that the relevance of non-formal Com-mission powers 'has not been analyzed systematically'. To learn more about the generation of policy alternatives and the question of how the processing of issues structures policy choices, however, the analysis needs to move beyond formal inter-institutional analysis and begin to incorporate insights into the organisational underpinnings and the system of interest representation in EU legislative policy making.

Organisational foundations and bureaucratic politics

If the EU fails to exhibit uniform and stable patterns of policy making across policy issues, viewing policy dynamics in their immediate organisational contexts might offer additional insights. Exploring Schattschneider's (1960: 71) dictum that some conflicts are organised into politics while others are organised out, Egeberg (2004) outlines a research perspective that seeks to 'unpack the basic organisational characteristics of the institutions within which individuals interact' (2004: 201) and asks how they structure the ways policy issues are perceived and processed. As discussed above, the Commission holds the exclusive right of legislative initiative under most articles of the treaty.

Its prominent position during the policy initiation stage has thus led researchers to assume that the Commission's internal organisation and distribution of responsibilities greatly affects how policy issues enter the formal EU agenda and, by extensions, how issues are packaged and prepared for decision making. Chiefly among the factors traditionally given attention from this perspective are the Commissioners' portfolios and the largely corresponding organisation of the Commission's administrative services, the directorates-general. Cram's (1994) work on the Commission as a 'multi-organisation' has already drawn considerable attention to the fact that the Commission's different administrative branches often perceive problems, process information and pursue programmes independently. Frequently, portfolio organisation yields only partial clarity, however, since Commission services and Commissioners' portfolios are sometimes organised vertically (such as covering one specific sector like agriculture), and sometimes horizontally (such as covering issues of public health across sectors). When policy issues cannot be unambiguously classified as falling under the purview of a single member of the Commission, which is frequently the case, assigning joint responsibilities for the drafting of a legislative proposal often ensures that multiple policy perceptions remain actively in play even at the outset of the EU legislative process, practically activating multiple, competing organisational logics right from the start.

In addition to interdepartmental conflicts, the Commission has also been found to harbour a 'variety of contradictory organisational logics' as tensions between its bureaucratic and political functions surface frequently (Christiansen 1997: 87). Few Commission decisions can be clearly defined as either political or bureaucratic, and the Commission's organisational structure equally reflects both logics, routinely pitting Commissioners and their political staff against the Commission's administrative branches. Further complicating the picture, organisational tensions also prevail at the level of individual agency inside the Commission. The Commissioners themselves have been found to shift allegiances between the Commission's institutional interests as a collective actor and political rationalities that are based on their policy portfolio, country or origin, or party membership (Egeberg 2006). Employing different research strategies to address the same puzzle, Wonka's (2008: 1159) study 'supports Egeberg's conclusion that Commissioners' behaviour is driven by multiple factors' with only slight variation in the importance attributed to the different organisational logics at work. Even seconded national experts, those members of national bureaucracies sent to Brussels for limited time periods, fail to exhibit clear and persistent role orientations while serving at the Commission. Instead, they were found to have 'a triangular behavioural pattern that is dominated by departmental, epistemic and supranational dynamics', as Trondal (2007a: 173) found in a study that largely confirmed results from a similarly research project ten years earlier (Egeberg 1996). If anything, portfolio organisation seems vital (e.g. Trondal 2007b: 963), but its influence is clearly variable and the functional organisation of responsibilities inside the Commission is no reliable institutional determinant of policy perception or political behaviour.

Little indicates that the other EU institutions produce less ambiguous organisational logics. The same mixed findings instead appear to hold analogously for the Parliament, where EU deputies not only meet in the assembly, in committees, sub-committees and party groups, but also as national party delegations, with national party meetings often preceding the meetings of the full political group (Raunio 2000: 239). At the same time, 'conflict between the committees over policy competence' has become more common (Burns 2006: 237, see also Neuhold 2001). Far from imposing a single uniform logic of organisational behaviour, Whitaker (2005) points out that the rules and procedures that guide committee activity in the European Parliament are instead designed to ensure that multiple political logics of representation are at work, be they functional, ideological or national, and that none of them seems to prevail consistently. Not surprisingly, members of the European Parliament have been shown to 'interpret their own representative roles broadly, believing that they have important responsibilities towards multiple constituencies' (Scully 2003: 285). According to some accounts, deputies' role orientations were found to be shifting between supranational, national and party roles. A different study finds some support for the predominance of traditional ideological divides, but instantly cautions that 'national divisions, north/south divides and pro- or anti-EU stances surface *depending upon the issues at hand'* (Mather 2001: 193, emphasis added).

In her resulting criticism of the European Parliament's unclear representational function and the kaleidoscope of viable political rationalities, Mather (2001: 192) eventually concludes that 'it is left to the individual representative to determine her/his stance on every issue'.

Far from supporting any straightforward interpretation concerning dominant political role behaviour or prevailing logics of representation, empirical research on the organisational behaviour of EU supranational actors increasing points towards complex and contingent patterns that vary according to the issues at stake and serve to highlight the EU institutional actors' remarkably capability for 'decomposing' (Kassim 1994: 24) into contested organisational arenas. Despite the fact that recent research has made impressive strides in unpacking the institutions of the EU from an organisational perspective, the unresolved complexities continue to impinge on our ability to formulate predictive theory of the EU policy process without knowledge of the policy issues in play. While it is less clear how and when any particular role identification of an individual EU actor will be trumped by another more salient organisational or representational frame of reference and give rise to a specific logic of political behaviour, research abounds that tells us that shifts happen all the time and that the political role identification of supranational actors in the EU remains astonishingly ambivalent. In sum, it appears true for most actors on most levels in the EU that 'representation involves balancing multiple competing roles in different situations at different times' (Trondal 2008: 447). Given the highly variant effects of EU political organisation on policy choices, the institutional level of analysis alone thus remains insufficient to predict how issues and interests will play out at the supranational level.

Policy networks and organised interests

A more encompassing view of the structure of political representation in the EU policy process moves beyond the inter- and intrainstitutional focus and looks at interest intermediation more broadly. Among theories of EU interest politics, policy network analysis has long claimed a privileged position. Policy network analysis in EU studies has its origin in the context of the multi-level governance literature that studies the interdependence of political decision making at the sub-national, national, and EU level. Despite these theoretical roots, however, the study of policy networks has increasingly informed EU policy research with a more exclusive focus on the supranational level (Pollack 2005: 382–385). This trend has not replaced the study of vertical policy networks and EU 'governance' (Hix 1998), but it reflects the growing realisation 'that the policy networks that matter most in EU decision-making are more horizontal in structure' (Peterson 2004: 132). Nationally organised interests are increasingly forming interest representations of various types at the EU level and seek access to the Brussels-based institutions after learning that domestic strategies of interest organisation often render little influence too late. Especially with respect to the early stages of the EU policy process, EU level interest organisations have been found to play a more significant role than interests organised at other levels (Eising 2004: 236; see also Bouwen 2004 and Coen 1998). This trend has led some observers to diagnose an 'erosion of state control over domestic interest groups' (Richardson 2000: 1015), while the accompanying interest reorganisation at the EU level 'weakens one of the key features of the glue holding national policy communities together.' Despite the increasing focus on the supranational arena, the direction of the main causal argument explored in the policy network literature has remained unchanged. It typically investigates how interest and policy perceptions originate outside the political institutions of the EU, and how pressures are subsequently directed at the supranational actors and institutions involved in a policy decision. The reverse perspective, addressing questions of how the supranational arena in turn affects the formation and representation of interests in EU politics, is less well developed.

That the direction of influence goes both ways, however, becomes partially clear from studies that look at the size and durability of EU policy networks. While several approaches in this line of research, such as the analysis of advocacy coalitions (Sabatier 1998: 103–104), are explicit about the long-term stability of policy networks as a precondition for their success, several studies have raised doubts as to whether the EU policy making system is really as hospitable to organised interests as casual observers regularly seem to imply. Occurrences of policy networks that dominate EU policy making over extended periods of time have instead been found to be rare. 'The characteristics of EU policy making', Richardson (2000: 1021) argues, 'do not seem conducive to the systematic emergence of traditional policy community politics or to stable policy networks as a system of governance at the EU level. Moreover, the characteristics of EU policy making help to undermine stabilized patterns at the national level.' Concluding a detailed analysis of the application of policy network analysis in EU research and the obstacles this type of interest organisation faces in the EU political system, Kassim (1994: 21) equally points out that 'there is no compelling evidence to suggest that durable patterns of interest intermediation are materialising, while the location where decisions are taken in the EU may change unpredictably, according to strategic choices made by institutional actors'. More recent assessments of EU interest politics give no indication that this has changed. Peterson (2004: 127), for example, describes policy making in Brussels as 'too fluid, uncertain, and over-populated with an enormously diverse collection of interests for stable networks to exist or persist'. Partially because of the predominance of regulatory policy making in the EU and its effects on the political organisation of interests, the import of more centralised and cohesive forms of interest intermediation has predominantly failed. Since the 'costs and benefits of EU integration are likely to be distributed unequally across the national and subnational constituencies of existing associations', Traxler and Schmitter (1995: 192) explain, 'the task of defining member interests and elaborating legitimate common demands is bound to become more difficult'. Eventually, these authors arrive at the conclusion that 'increasing pluralism in both the mode of intermediation and the form of interest organization' at the EU level is unavoidable (1995: 213). Surveying the field of EU interest politics more broadly, Woll (2006: 459) equally notes that 'most examinations conclude that effective collective action at the supranational level is difficult even for groups of large and powerful actors such as firms'. As the consequences of collective action problems in a multi-level system become more apparent, she goes on to argue, 'studies have increasingly focused on new groups and ad hoc alliances', many of which are limited to the sub-field level of policy or even single policy issues (see also Pijnenburg 1998).

In this context, studies of EU institutions as incentive and opportunity structures have gained in importance (e.g. Princen and Kerremans 2008; Cram 1998). Richardson (2000: 1018) elaborates this point of view in a study of EU policy research that investigates how successful policy communities thrive to adapt to prevailing problem perceptions at the EU level, rather than attempting to impose their own views and interests. Radaelli (1995a) similarly points out how a focus on knowledge and expertise with an only peripheral interest in political power structures makes knowledge-based approaches of network analysis particularly prone to ignoring the fact that expertise only enters the EU policy process if it is compatible with the prevailing problem perceptions. While the policy network literature often formulates accounts of policy making that conceptualise a de-politicisation of democratic decision processes and place special emphasis on the role of knowledge instead, students of EU interest politics warn that 'European policy making is not a depoliticised process wherein the exchange of functional expertise and information stands central' (Beyers and Kerremans 2004: 1146). Pushing argument concerning the influence of supranational institutions on EU interest intermediation even further, Mahoney's (2004) research highlights the influence of EU institutional actors on interest organisation in Brussels through direct subsidies, or by way of manipulating the scope of EU policy issues and the formal arenas available to discuss them. Beyond the lesson that the successful representation of interests at the European level appears less dependent on the pre-existing modes of organisation at the national levels or elsewhere, and more dependent on the adjustment to the structure of the issues, allocations of responsibility and organisational self-interests of EU institutional actors as well as the generally unaccommodating structure of the EU policy process, these factors themselves often remain difficult to pin down.

As recurring patterns and processes of EU interest intermediation remain difficult to identify at aggregate levels of analysis, current research on EU interest politics has echoed calls for a more explicit theoretical focus on the respective nature of specific policy issues on the EU agenda (see also Mahoney and Baumgartner 2008: 1267–1268). 'All of these studies', Woll (2006: 460) sums up, 'share the view that EU lobbying cannot be understood without looking at the institutions and policy context in which the groups are trying to act'. Similarly, Eising (2007: 356) argues that is remains 'difficult to identify general patterns of interest intermediation in the EU because the impact of many factors is conditioned by the segmented EU institutional context'. These theoretical explorations thus largely confirm the main points Kassim (1994: 20) made a decade earlier. 'The fragmentation of the policy process has the consequence', he concludes in his study of EU interest intermediation, 'that the difference between policy issues may be more significant than any similarities at the sectoral, or even subsectoral, level'. As a result, Kassim advises that in EU policy research, it probably 'make more sense to focus on individual decisions or issues rather than to address broader units of analysis' (1994: 21).

The role of policy issues

Harcourt's (1998) study of EU media ownership regulation exemplifies how a research strategy that traces political issues back to the early stages of policy formulation calls standard conceptualisations of EU political actors and their interactions into question and 'enables the researcher to discover actors empirically' (1998: 387). While she acknowledges that the Commission acts as a unitary political decision making body at crucial points in the legislative process, her research perspective also calls for the analysis of the 'Commission as a political system' (1998: 371). She shows how various directorates-general and societal interests compete throughout the policy formulation process over the right to define issues and claim legislative responsibility for them. According to Harcourt (1998: 370), policy frames are at the centre of the political conflict because frames 'empower certain actors over other actors'.

Lenschow and Zito (1998) similarly emphasise that the effects of policy framing and reframing need to be addressed in the context of contested institutional competences and observe how these 'political struggles… between EEC institutions have overshadowed substantive argumentation' (1998: 428). Confirming the centrality of framing disputes, Dudley and Richardson (1999: 228) equally note the absence of any attempt by EU political actors to bridge perceptional differences and avoid political conflict over policy decisions. While this research shows how institutionally embedded policy frames work to 'bias the policy discourse' and 'constrain the range of options decision makers perceive as available' (Lenschow and Zito

1998: 419), these authors also point out that in the EU many of the organisational and procedural factors of their model are in flux. Looking at the cross-pillar issue of the European defence industry and equipment market, Mörth (2000) provides a particularly insightful example of an institutionally contested policy area from this vantage point. She sets out by analysing how the fragmented organisational structure of the Commission is prone to generate competing policy frames and then proceeds to investigate how issues are framed and reframed as they move from one DG to another (2000: 176–177). By redefining the issue at stake, Mörth argues, the Commission copes with intra-organisational conflicts while, at the same time, it crafts support and pre-structures policy formation in an area of contested EU jurisdiction and weak Commission competences.

In light of these and similar findings, the EU has sometimes been called an 'agenda-setter's paradise' (e.g. Peters 2001: 88) in which issues can be introduced and pursued at different levels and across different institutional venues. Yet while frame competition inside the Commission in fact helps to draw diverse and sometimes conflicting issues and interests into the EU decision making process, students of EU policy framing also note that supranational institutions, most notably the Commission, just as frequently take up issues even when no initial expression of demand for new policy is evident. Reviewing the legislative process that led to the liberalisation of the EU electricity market, Nylander (2001: 307) for example posits: 'Roughly speaking, the Commission started out in the late 1980s without consulting anyone.' The Commission is hence not only selective in the way it engages interests in policy deliberation, political interests are also often found to form only in response to specific policy initiatives, rather than the other way around. Existing work on EU policy framing specifically indicates that the Commission routinely and systematically sets out to construct political support for policies, even against overt political resistance. Wendon (1998) offers an example of how the European Commission manipulates the formation of interests in the EU by changing the perception of policy issues and by providing for new institutional venues to process its initiatives. In the case of EU social policy, the Commission initially failed to formulate an agenda on which the key actors could agree. It then reframed the issue 'by shifting the debate... from technical issues to ethical considerations of social justice' (1998: 345) and subsequently presenting 'social policy as a productive dimension' (1998: 350). In doing so, the Commission managed to draw new actors into the debate and at the same time facilitated the creation of new policy arenas in which its policy initiatives were more favourably received. Very similar lines of argumentation are advanced by Smyrl (1998) in his discussion of the case of the Integrated Mediterranean Programme. He concludes that it would have been impossible to predict eventual policy choices in the Council based on member states' preferences at the outset. Instead, a 'focus on the internal process of policy conception brings the light of analysis to bear on the Commission's role in specifying and at times reconceptualizing the alternatives among which Member States will ultimately choose' (1998: 96; see also Bauer 2002). In short, 'states and interest groups are mobilized by means of policy frames' (Nylander 2001: 292). If the flow and structure of policy issues is constitutive of political dynamics in the EU, however, the theoretical exploration of EU policy research needs to shift focus. The next section summarises how framing research addresses these issues as an analytical perspective on the EU policy process and previews the main findings of this study.

Preview of the argument

Identifying stable patterns of policy contestation in the EU has proven difficult. In this policy making system, even entrenched political interests can find themselves easily sidelined as the legislative process unfolds. Correspondingly, policy issues have emerged as a key unit of theoretical exploration. How interests form and organise in EU policy making, how they find expression in policy debates and whose demands are processed varies according to the faming of the issues at stake. Rather than asking how existing cleavages in EU politics, established patterns of interest group representation, or the institutional organisation of the EU determine the processing of EU policy issues, the reverse logic of influence appears at least as critical. The way in which issues are perceived, packaged and processed systematically affects how policy conflicts form around the issues, how interests mobilise and restructure at the supranational level, and which of the conflicting organisational and representational logics of EU decision making shape the lines of both intra- and inter-institutional competition. As Radaelli (1995b: 158) notes, EU initiatives that fail under one policy frame can become feasible under a different frame. If EU policy choices reflect dynamics triggered by competing and selective perceptions of political issues, however, supranational policy making must be conceptualised at a fairly disaggregate level to offer a 'better specification of actor preferences... and a nuanced and empirically more accurate picture of the relationship between these actors', two qualities Pollack (1997a: 131) identifies as critically lacking in theoretical models of EU legislative politics. This type of argument would seek to accommodate more detailed accounts of the various organisational logics and levels of EU politics and focus on how they come to influence the emergence and containment of EU policy dynamics. Conflicts often form along only a few dimensions, as the spatial literature argues, but these dimensions are not external to the policy process. By placing policy level contestation and its effects on the political decision making process of the EU at the centre of the investigation, framing analysis speaks directly to these issues. The question of how the structure and flow of policy issues drive this process and shape the ways in which policy choices unfold in the EU therefore requires more systematic attention.

Complementing more traditional studies of EU agenda setting, this book shows that the way in which issues are framed is just as important for a better understanding of EU politics as the question of what issues are placed on the EU agenda to begin with. Drawing on the empirical investigation of legislative initiatives from over two decades of EU biotechnology policy making, this book investigates how shifting perceptions of the legislative policy issues on the EU agenda allow actors in supranational politics to shore up support and marginalise political opposition. The book finds that the framing of policy issues systematically affects how the complex and fragmented EU political decision making process involves or excludes different sets of actors and interests from the diverse political constituencies of the Union. Especially the European Commission is shown to define and redefine the issues at stake almost every step of the way. The empirical evidence further illustrates that framing decisions inside the Commission are often highly controversial political choices. The findings hence corroborate the claim that issue frames have systematic causal effects on subsequent policy dynamics. In particular, the empirical investigation confirms that the Commission's attempts to frame and reframe policy initiatives substantially shape political interest formation at the

European level. Yet the longitudinal perspective adopted in this study also reveals how the structuring and restructuring of the biotechnology policy debate led to the increasing politicisation of the EU decision making process. In democratic politics, policy framing rarely goes uncontested. The repeated attempts at framing and reframing the issues on the EU agenda left the choice over biotechnology policy politically charged. The Commission eventually lost control over the resulting policy dynamics, and the revision of the EU regulatory framework ran counter to the Commission's original policy objectives. The comparative analysis of legislative case studies and the longitudinal perspective of this book reveal that the possibilities to frame EU policy issues in such a way as to facilitate stable majority support can vary considerably. In sum, the framing perspective developed in this book provides fresh insights into how policy framing affects political conflict and competition in the EU. How policy issues are defined greatly affects the EU political decision making process and subsequent policy choices. This book illustrates that the role of policy framing in structuring EU policy conflicts can at times be substantial. But the framing perspective also shows that the possibilities to sway other actors in EU legislative policy making are constrained and reveals that policy reframing in the EU can result in persistent periods of political volatility.

Strategies of inquiry

Case selection

The framing perspective advanced in this book selects policy issues as the main unit of empirical analysis. The individual case studies go back to the earliest biotechnology policy initiatives at the EU level in the 1980s and cover the following two or more decades of policy expansion and revision. Each of the four laws addressed in the empirical analysis underwent substantial review and reframing before the EU biotechnology policy framework reached a more stable state around the year 2003. During the time period under analysis, each of these legislative initiatives was taken up for legislative action during two different framing periods. This scheme hence produces eight empirical observations of central importance:

- Directive 90/219 on contained use: adopted 1990 and amended 1998

- Directive 90/220 on deliberate release: adopted 1990 and repealed 2001
- Regulation 258/97 on novel food: adopted 1997 and replaced 2003
- Directive on legal protection of biotechnological inventions ('Biotechnology Patents Directive'): vetoed in 1995 and adopted 1998 as Directive 98/44

The case selection further allows for variation on the outcome variable in the sense that not all policy initiatives succeeded – far from it. The cases discussed here include one of the first vetoes of an EU law by the European Parliament and a moratorium on the implementation of the existing EU regulatory scheme by the Council. Following impressive legislative victories during the early days of EU biotechnology policy making, the subsequent reframing of the issues ended in havoc and left the Commission struggling to gain control of the policy developments in the late 1990s.

Strategies of causal analysis

The empirical investigation pursues a combination of research strategies to increase the possibilities of causal inference. They include both cross-case comparison and within-case analysis to compensate for the limits of each method in the context of the empirical investigation. The primary interest of this book is to study the effects of policy framing on policy choices. The most obvious test of the framing argument therefore observes how framing and reframing affect legislative votes. One important criterion to assess the plausibility of underlying causality is congruity (George and Bennett 2005: 183): causal and outcome variables are congruent in the nature (e.g. magnitude, direction, duration) of their variation. As George and Bennett (2005: 183) point out however, to assume that dramatic variation in the causal variable leads to dramatic effects (and the reverse logic, according to which dramatic effects can only be caused by dramatic variation) is often a fallacy. Indeed, the claim of framing approaches is precisely that small variation can at times bring about dramatic effects on the systemic level of policy choice. Accordingly, congruity tests need to be informed by the theoretical elaboration of causal mechanisms and prove informative only in combination with process tracing. The study of causal mechanisms focuses the attention on the underlying processes of policy making (e.g. George and Bennett 2005: 131–149; Mahoney 2001: 577–581; Hedström and Swedberg 1996). The soundness of this type of causal inference hinges primarily on the incessant interlocking of observable processes as discussed in Chapter 2. Compared to alternative research strategies, process tracing pursues a middle way between a narrow focus on often little understood causes and their expected effects on the one hand, and indiscriminate contextualisation of causal factors on the other hand. In the context of what Mahoney (1999: 1168) refers to as the 'overarching macrocausal argument', this type of analysis is 'used to evaluate expectations by establishing at a more disaggregate level whether posited causal mechanisms plausibly link hypothesized explanatory variables with an outcome' (see also Gerring 2005: 178–179). In order to do so, the different stages of the political process must themselves be described as interlocking causal relations that can be observed independently. Mahoney (2000: 414) especially emphasises the importance of process tracing for studies 'in which explanatory and outcome variables are separated by long periods of time'. Legislative initiatives at the EU level often take years before they are passed into law, making process tracing a particularly adequate research strategy.

While causal mechanisms identified in a single case can provide valuable insights, Mayntz (2004: 241) insists that the notion of causal mechanisms implies processes that are systematic and recurrent (see also Elster 1998: 45). The empirical analysis of four legislative initiatives over extended periods of framing and reframing easily allows for the investigation of matching patterns of policy making. Furthermore, 'pattern matching helps narrow the range of potential explanations by offering an additional means of eliminating variables' (Mahoney 2000: 411). Over the course of the empirical analysis, for example, the insights into framing effects gained from the study of one case can thus be used to further structure and focus the subsequent empirical analysis, leading to cumulative insights (George and Bennett 2005: 67) and eventually helping to refine the theoretical framework. When traditional cross-case comparison is hampered, Mahoney (2000: 415) consequently finds that to 'compare event sequences across cases to determine whether cases can reasonably be seen as following aggregate causal patterns at a more fine-grained level' constitutes an essential part of the research design (see also Mahoney 1999: 1165–1167).

Sources

The empirical material used in this study broadly falls into two categories: primary sources that are directly related to the EU institutions and their activities, and secondary sources that provide the necessary context in which to interpret the legislative acts. The first group of documents primarily encompasses the EU directives, regulations, and decisions themselves, from their initial form as Commission proposal to the final legislative language. In addition, the Biotechnology Documentation Centre (BIODOC), run by DG Research, was consulted repeatedly during the course of this research. This archive contains a unique collection of printed documents on EU biotechnology from various sources, mainly from inside the Commission, but also including public documentation. Especially for the reconstruction of the early policy making years, the BIODOC archive proved crucially important. Secondary sources such as the *European Voice* (starting with the first edition in 1995), *Agence Europe*, the *Bulletin of the EU*, and the *European Information Service* (EIS) press releases were consulted systematically as part of the reconstruction of the political processes that linked the initial policy initiatives to the final act of voting. Combined, these news sources provide detailed, often weekly or even daily reports that made it possible to trace policy developments with great depth and accuracy. Interviews with senior members of the Commission and Parliament staff as well as key interest group representatives were conducted to gain a more detailed understanding of the background, and to inquire about facets of the political process that could not be deduced from the consultation of written publications.

Structure of the book

Chapter 2 explores the framing perspective in political analysis. Beginning with the classic literatures on agenda setting and political problem definition, it discusses how recent advances in policy framing research systematically tie the definition of issues during the phase of policy formulation to the political dynamics that shape legislative policy making and outcomes. The conceptual and theoretical discussion focuses on three distinct functions of policy framing in political decision making: the reduction of issue complexity, the formation and organisation of collective interests, and the structuring of political conflict and competition.

In Chapters 3, 4 and 5, four legislative initiatives from the field of EU biotechnology policy are traced over more than two decades. Over this time period, EU biotechnology policy turned from a regulatory nether land into one of the most volatile and bitterly contested policy conflicts in the European Union. The empirical analysis covers the creation of this policy field at the EU level and the European Commission's subsequent attempts at reframing the issues in pursuit of a new policy agenda. The resulting politicisation of the EU policy choices cumulates in one of the first vetoes issued by the European Parliament and a Council moratorium on the implementation of the existing legislative framework. The empirical investigation of these political dynamics highlights how the conflict over the framing of the policy issues restructured the policy field and eventually led to the adoption of a revised and expanded regulatory framework at the EU level, contrary to the Commission's original policy objectives.

Chapter 6 reassesses the framing perspective in light of the empirical evidence and asks what lessons can be drawn concerning the scope and nature of policy framing arguments in the European Union. The first part of the chapter reviews how the architecture of the policy conflict was transformed over the course of more than two decades and how it recast political alliances in the process. Focusing on successes as well as failures of policy framing, the analysis highlights how EU actors compete over ways to shape the conflict space at the supranational level. The discussion concludes that the EU policy making system is highly volatile. Lacking a stable political conflict space that structures policy debates over time and across issues, the EU policy system leaves supranational actors prone to contest the issues on the agenda along fluctuating dimensions. As a result, the consolidation of conflicts in EU biotechnology policy making had a seemingly idiosyncratic dynamic. The chapter shows that framing research contributes an analytical perspective that yields systematic insights into the underlying processes at work and provides insights that are both theoretically consistent and empirically more accurate than contending theoretical lenses.

The final chapter summarises the main findings and places them in the larger context of both policy research and the study of EU politics.

2
Policy Framing Analysis

E. E. Schattschneider's (1957: 937, 1960: 66) aphoristic definition of policy alternatives as the supreme instrument of power in politics continues to challenge research in political science. His work helped to establish the notion that agenda setting structures political choices (see Bachrach and Baratz 1962, 1963; Cobb and Elder 1971; Baumgartner 2001). Yet the majority of the research that followed in his footsteps abandoned Schattschneider's comprehensive view of politics as a volatile process of choice and focused more narrowly on the early stages of policy making. An analytical perspective designed to guide the systematic investigation of how the definition of policy issues affects subsequent political dynamics did not emerge. Recent advances in the study of political rationality (Simon 1983, 1985, 1987, 1995; Jones 1994a, 1994b) have revitalised political science interest in this theoretical puzzle. Framing arguments were introduced to a wider social science audience especially through the work of cognitive psychologists Tversky and Kahneman (1981, 1986; see also Quattrone and Tversky 1988). Their research addresses the effects of the representation of alternatives on the evocation of interests in the process of decision making. The main argument of this literature is that every decision can be framed in different ways and that choices systematically vary in response to the reframing of the issues. From a decision-theoretic perspective, frames therefore constitute fundamental factors in explanations of political choices. But policy framing research must also account for influences beyond the individual level of decision making. The institutional structure of policy making systems has its own effects on how political problems are perceived and how issues are processed (Jones 1994b; Baumgartner and Jones 1991).

Policy framing analysis must therefore take account of the way in which political institutions delineate policy jurisdictions and structure policy debates. If the reframing of policy issues gives rise to turf battles between competing institutional arenas of government, for example, the results must be expected to transform policy conflicts and generate contestation on entirely new levels of the political system. Finally, frames are themselves subject to political play. Policy makers advocate competing frames strategically to restructure the political playing field and sway votes in their favour (Riker 1986). For students of the policy process, straightforward lessons from the literatures on framing in decision making analysis and organisation theory are thus not always immediately forthcoming. At the same time, framing arguments of the policy process require a partial rethinking of central arguments in traditional policy research and sometimes recast established causal relationships.

At the most fundamental level, the policy framing perspective contends that the definition of policy issues affects the processing of political ideas and demands and their subsequent expression in policy choices. Issues drive the policy process. The flow and structure of policy issues affect the political alignment of actors and the conflict and consolidation of interests in the policy process. Policy issues are hence not politically neutral. This chapter asks how research premised on these ideas can provide a framework for policy analysis. It shows that in combination with classic arguments in policy theory, framing analysis can greatly help to explore and explain dynamics of political decision making and offer new perspectives on the policy process. In advancing this analytical framework, the chapter first discusses conceptual tenets of framing research more generally and then goes on to focus more closely on the question of how framing analysis can play a role in explaining policy dynamics. Based on the theoretical discussion, the chapter then turns to address research strategies that can guide the empirical investigation of EU policy framing and shed new light on the internal politics of the European Union.

Theoretical origins

Agenda setting

With his famous conclusion that 'the definition of the alternatives is the supreme instrument of power... because the definition of the alternatives is the choice of conflicts, and the choice of conflicts allocates power', Schattschneider (1957: 937) paved the way for the study of agenda setting in political science. Boldly reconceiving politics as the use of conflict to govern (Schattschneider 1957: 935, 1960: 69), research in this tradition addresses the political dynamics lying behind the overt workings of democratic government and criticises pluralist models of democracy for 'exhibiting relatively little formal concern with the scope of participation and influence in the determination of political alternatives' (Cobb and Elder 1971: 896). Instead of taking political interests at face value and interpreting them as reflecting the underlying issues in a straightforward way, students of agenda setting argue that the policy making process is prone to bias. Accordingly, an adequate and critical assessment of any political system must start at the outset by asking which choices the political system facilitates, and which interests and alternatives are excluded from the political process. In other words, any given policy choice must be analysed in terms of the biases that created it.

Following the rise of agenda setting research, policy processes that had until then been labelled 'pre-political' became the name of the game. Schattschneider maintains that what happens in politics ultimately depends on the way in which the actors are divided. Yet the factions they form and the positions they take on the issues are not fixed or given. Instead, they depend on 'which of a multitude of possible conflicts gains the dominant position' (Schattschneider 1960: 60) at a certain time. To conceptualise the ways in which conflicts play out in the process of political decision making, Schattschneider analyses them in terms of their intensity, visibility, scope and direction, and argues that manipulation of these factors transforms them into instruments of political strategy. This theoretical lens focuses on how the scope of a conflict develops strategic implications when advocates of a minority position redefine an issue so as to expand the relevant public and attract more social and political actors. Schattschneider refers to this strategy as the expansion (or socialisation) of conflict. Through conflict expansion, one political camp can gain strength by activating more contestants and resources. Most importantly, however, every expansion of scope brings about a shift in the direction of the conflict. 'Every change in the scope of conflict has a bias', Schattschneider (1960: 4, see also 1957: 942) stresses. As new contestants enter a debate, the lines of conflict shift and tilt, new alliances become possible, previously aligned actors split, and opportunities for change arise. Schattschneider regards this effect, which he terms the displacement of conflict, as the most consequential bias of the democratic political process. The bottom line of this argument became known as the 'two faces of power' (Bachrach and Baratz 1962, see Baumgartner 2001 for a concise summary). According to this perspective, the issues which enter political agendas, and the alternative responses to them that are considered, are the result of factors that operate before decisions are taken and votes cast in political institutions. Every political system, the argument runs, encompasses choices that never have to be faced. Some interests are prevented from forming and some policy alternatives are eliminated without ever being considered. Such cases of 'nondecisions', Bachrach and Baratz (1963: 641) argue, result from structural constraints that 'effectively prevent certain grievances from developing into full-fledged issues which call for decision.' While 'nondecisions' them-selves naturally defy observation, the authors hold that they result from political processes and structural biases that fall into the realm of political analysis. The 'two faces of power' argument finds a more limited but academically influential expression in Cobb and Elder's (1971) distinction between 'systemic' and 'institutional' political agendas. The first refers broadly to all issues under consideration in a polity at a given time. The second refers to the relatively few issues that the institutions of government take up and process for decision making. Guided by the question of how issues shift from one agenda to the other, the authors focus on the processes through which issues are created. As a result of this research, Cobb and Elder (1971: 903, 905) conclude that the 'pre-political, or at least pre-decisional, pro-cesses are often of the most critical importance in determining which issues and alternatives are to be considered by the polity and which choices will probably be made.' The influence of pre-decisional pro-cesses is thus not limited to the gate-keeping function discussed in the work of Bachrach and Baratz (1962, 1963). Instead, Cobb and Elder (1971) stress that bias is a universal feature of the political process and that the effects of this necessarily selective process of decision making must assume much more subtle forms than 'nondecisions'.

With a focus on the political bias in policy making came a renewed interest in the role of political elites and political entrepreneurs. As a result, the pluralist tradition of thinking was shaken up entirely. Cobb,

Ross and Ross (1976: 132) note that instead of giving voice to existing grievances, the political mobilisation of non-governmental actors by policy makers more often results from the fact that outright coercion is deemed 'inappropriate, impractical, or simply too expensive.' Far from constituting the sovereign source of political empowerment, the public is portrayed as divided into identification groups, attention groups and the mass public. These groups exhibit different degrees of attachment to any given political issue. By choosing whom to engage and whom to keep out of the political deliberation process, their argument concludes, policy makers can shore up support, deliberately create friction and mould the political process towards predefined ends. In one of the first systematic studies of the transition of political issues from the systemic to the institutional agenda, Cobb, Ross and Ross (1976) identify three distinct paths. Tellingly, two of the three paths describe top-down policy formulation that originates in the political as opposed to the public realm. One model describes policy making with no extension of the issue to the public at all. From this perspective, politics is more concerned with generating the power to govern than with the redress of societal grievances. The notion of the 'autonomy of the political' (Mair 1997: 950) is central to this perspective. Politics is seen as developing its own momentum and evolving according to an inherent logic, which often bears little resemblance to more common notions of problem solving. How the inherent dynamics of politics play out must instead be traced back to the framing and reframing of political issues.

Issue definition

According to agenda setting research, political strategy is thus perceived to gain substantial leverage when it manipulates the definition of political issues. But clearly, not all issues are inherently capable of producing the same kinds of conflict or setting off the same kinds of dynamics of political interaction. While Schattschneider theorises that the displacement and enlargement of conflicts always goes hand in hand with issue redefinition, the reverse must not always be true. The possibilities for managing politics through the framing of policy choices are not unbound. Instead, and analytically much more intriguing, some issue characteristics appear to trigger distinct political dynamics in systematic ways. The search for theoretical angles that can capture these causal relations, however, has proven difficult. In-depth case research of single political issues such as Nelson's (1984) classic study of child abuse provides convincing evidence of the importance of the definition of political issues. Other major contributions to this growing literature cover a wide variety of approaches to the question, ranging from the analysis of causal stories in policy making (Stone 1989) to the impact of instrumental versus expressive rhetoric (Rochefort and Cobb 1994: 159–181). But the theoretical task of identifying systematic effects has proven to be excruciatingly difficult. As a result, much of the literature on political problem definition departed from Schattschneider's original focus on the role of issues as drivers of the political process and moved on to study the social construction of problems and their expression in policy making (see Rochefort and Cobb 1993: 57). One major drawback of this approach is that arguments based on socially constructed issue perceptions and their influence on policy choices cannot account for variation on either side of the argument, thus defying the notion of the political process as a consequential and biased force in politics rather than illuminating it. As a result, the bulk of this literature has not moved significantly beyond the 'two faces of power' argument and the analysis of agenda denial (see Cobb and Ross 1997). The link between the representation of political issues and the representation of interests in the policy making process remains partially unclear. Over the past two decades, however, a growing literature on framing in political science has contributed a distinct theoretical body of research and found new ways to address Schattschneider's puzzle.

Political rationality

Focusing on more limited choice situations reflected in voting records and laboratory experiments, recent advances in decision making analysis have made great progress in formulating links between issue representation and choice. Especially the research of cognitive psychologists Tversky and Kahneman (1981, 1986) popularised the notion that 'alternative descriptions of a decision problem often give rise to different preferences' (1986: S251). Their subjects, for example, change preferences for medical treatments depending on whether outcomes are described in terms of 'deaths avoided' or 'lives saved', and they change preferences for public policies depending on whether they are based on rates of employment or unemployment. The type of variation in choice behaviour described in these studies occurs despite the fact that 'alternative formulations of the problem convey the same information, and the problems differ from each other in no other way' (Quattrone and Tversky 1988: 735). Yet, contrary to the assumption of invariance that underlies rational theories of choice, alternative formulations of the issues produce predictable reversals in choice behaviour. Such findings raise serious questions about the ways in which decision makers reason. 'There is compelling evidence', Kahneman (1997: 123) concludes, 'that the maintenance of coherent beliefs and preferences is too demanding a task for limited minds'.

The argument that decision makers' computational capacities are less sophisticated and the task environment more complex than portrayed in standard theories of choice is most fully developed in the work of Herbert Simon. He argues that choices cannot be easily deduced from assumptions about the interests of strategic actors (Simon 1985: 297, 1986: S223, 1995: 49). Instead, the concept of rationality itself must be reformulated to contribute to our understanding of the processes of interest formation. The study of rationality as the process of choice leads Simon to a substantial reformulation of standard decision making analysis and opens up myriad ways of guiding and refining political analysis. Models of bounded rationality are premised on the simple truth that decision makers can only be rational in terms of what they are aware of. This argument is even more compelling if the models of choice assume that decision makers utilise information in sophisticated ways before choosing a course of action. In contrast to much stronger assumptions of universal rationality, bounded rational decision making theory presumes that decision makers' ability to reason is limited and that information has to pass the 'bottleneck of attention' (Simon 1985: 302). Since attention is scarce, much more so than information in most standard decision situations, information is processed selectively and successively. Incapable or reluctant to compare and evaluate attributes of a choice across multiple dimensions, individual decision makers, just as policy makers, struggle with ambiguities concerning the 'relevance, priority, clarity, coherence, and stability of goals' (March 1978: 595). One way to avoid confusion and trade-offs across different evaluative dimensions is to break down complex and interrelated issues into smaller, more manageable decisions. Yet decision making strategies that factorise choices in such a way 'work as intended only in a linear and decomposable world' (Jones 1994b:

49), a condition not frequently met in politics. Based on these assumptions, Simon formulates the 'design problem': how to define the contours and the nature of a choice. The design problem stems from two central decision making tasks, complexity reduction and goal formulation, and it involves the simultaneous search for the alternatives and the evaluative attributes of a choice (Jones 1999: 306; see also Jones 2001: 77, 274). If multiple facets of a problem interact, defining the issue is necessarily a highly discriminating process and prone to goal conflicts both at the level of individual choice, and even more so at the level of collective action (March 1994: 139–174). From this perspective, how policy makers perceive a decision problem is thus not only highly consequential. Problem designs are political choices in themselves.

Framing of choice

Instead of comparing attributes of a choice across multiple evaluative dimensions in ways that are systematic and stable over time, theories of framing in the tradition of research on bounded rationality hold that decision making is more volatile and can at times appear outright erratic. Despite this, such behaviour is neither irrational nor entirely unsystematic. The 'focus of attention', Simon (1987: 355) maintains, 'is a major determinant of the goals and values that will influence decision'. Consequently, to analyse (or control, for that matter) the mechanisms of attention direction is the key to explaining or manipulating complex choices. In this theoretical context, the concept of framing refers to the process of selecting and emphasising aspects of complex issues according to an overriding evaluative or analytical criterion.[1] 'Nowhere', Simon (1973: 275) contests, 'is the problem of attention management... of greater importance than in the political process'. When the focus of attention shifts, some facets of a problem are emphasised or deemphasised, some aspects of a decision are revealed and others ignored.

---

[1]Building consistently on the insights and terminology of Herbert Simon, Jones (2001: 105) arrives at a particularly parsimonious definition of framing as the 'phenomenon of directing attention to one attribute in a complex problem space'. The definition of framing used here selectively encompasses elements of those put forth by Entman (1993: 52), Gamson (1989: 157), and Weiss (1989: 118) to make explicit what the concept essentially entails.

As the representation of the issue changes, so does the perception of what is at stake, and the preferred solutions vary in response. Likewise, in the process of seeking out new alternatives, decision makers routinely come to reassess the relevance of their underlying interests. Which interests are evoked and how salient they appear thus hinges on the frame of reference. 'Incoming information', Jones (1994b: 238) elaborates, 'can either be put into existing frames... or can force a shift in evaluative focus'. In the latter case, 'policy issues are not just illuminated by information, they are framed by it. When issues are reframed, often through the highlighting of a previously ignored evaluative dimension, our basic understanding of an issue shifts' (Jones 1994b: 50).

From this theoretical angle, some issues of practical policy making appear in a new light. Instead of contesting arguments and facts, Baumgartner and Jones (1993: 107) write, 'it is generally more effective in a debate simply to shift the focus.' In stark contrast to argumentative policy analysis or theories of deliberation, framing analysis is essentially concerned with 'noncontradictory argumentation' (Jones 1994b: 182) – much in line with Schattschneider's emphasis on the role of conflict displacement – and places little or no emphasis on reasoning and persuasion. Framing strategies can operate unobtrusively and are therefore particularly useful to those political actors or institutions that cannot afford to act in an overtly partisan manner. Their impact on public debates, however, is often highly contagious. In his writings on heresthetics, Riker (1984, 1986) elaborates the political implications of framing (see Simon 1985: 302 for a rejoinder). He uses the term heresthetic to describe the manipulation of decision situations to make participants decide as the manipulator desires, despite an initial disinclination to do so. Framing analysis is a central concern of Riker's work on heresthetics. The 'manipulation of dimensions' of a choice, Riker (1986: 150–151) finds, 'is just about the most frequently attempted heresthetic device, one that politicians engage in a very large amount of the time... Most of the great shifts in political life result from introducing a new dimension.' Riker's work on heresthetics still serves as one of the most forceful exemplifications of framing effects, but it also highlights potential limits of the theoretical argument with particular clarity. The heresthetical interpretation of framing effects is based on the spatial model of voting, in which decision makers perceive of alternative political choices as positioned along evaluative dimensions. In many standard choice situations, voters' ideal points are portrayed as ordered along a single dominant evaluative dimension. In the case of simple majority voting and an odd number of voters, this situation has an equilibrium and hence a definitive result, the median. The heresthetic manipulation of dimensions gains power from the simple fact that '[this] neat equilibrium does not carry over to two dimensions' (Riker 1986: 145). Instead, the equilibrium is lost in most cases of multi-dimensional choice.[2] This, Riker (1986: 145) summarises, 'is why the complications of two dimensions (or more) guarantee that many outcomes are possible'. Accordingly, successful frame manipulators emphasise issue dimensions that separate the voters into 'a new majority-minority division' (Riker 1990: 51), which facilitates previously unobtainable political outcome. Framing, or heresthetic manipulation of issue definitions, thus derives its power from purposefully creating the instability of multi-dimensional choice. 'But heretheticians in the real world', Riker (1995: 34) warns, 'do not have exclusive access to the persons manipulated, nor do the manipulators have exclusive control over information, nor the exclusive right to formulate issues'. As a result, he laconically sums up his findings, 'real world heresthetical manipulation is sometimes successful, sometimes not.' Jones and Baumgartner (2002: 298) raise the same point when they note that multidimensionality in political decision making 'allows policy entrepreneurs to stress one attribute in a policy debate, but other participants are free to try to focus attention on a second, third, or even forth attribute of the issue'. In mass politics, moreover, dramatic focusing events can impose highly salient evaluative dimensions across policy fields (Simon 1987: 367; Birkland 1998).

In sum, Riker's spatial model of politics illustrates the effects of the framing on choice, but it cannot explain what renders a frame stable and hence consequential in the policy process. As decision makers are frequently left with 'contradictory and intermittent desires partially ordered but imperfectly reconciled' (March 1978: 598), a theory of stable and systematic framing effects in political analysis must go beyond the level of individual decision making behaviour.

---

[2]More precisely, Riker (1990: 50) argues that although an equilibrium may be theoretically well defined (like, for example, the multi-dimensional Plott equilibrium), these equilibria are virtually unheard of as real-world occurrences.

Simon's formulation of bounded rationality offers a much richer theoretical picture in this respect because it places shifting evaluative dimensions as moving parts of the explanatory model front and centre and links them back to attention dynamics. Building on these theoretical advances, framing theory in policy research needs convincing theoretical arguments at the organisational and institutional level of analysis to link issue-based conceptions of framing effects to systemic outcomes at the level of the policy making system (see also Baumgartner and Mahoney 2008). On what levels does policy framing operate? What are the mechanisms at play? Secondly, the scope of policy framing arguments has remained unclear. When does framing matter? How does counter-framing affect policy dynamics? Thirdly, the import of policy framing theory developed in the context of American political science runs the double risks of both overstating and underestimating the relevance of structural features peculiar to the case of the EU. How the unique set-up of the EU political system affects framing dynamics on the supranational level is a key concern to which the discussion will return in Chapter 6. In answering the first two sets of questions, the following sections expand on existing conceptual and theoretical insights and focus on three distinct levels or functions of policy framing in political decision making: the role of the organisational foundation of politics in the reduction of policy complexity, the creation and organisation of collective interests, and the structuring of political conflict and competition. All three functions are discussed in turn.

Complexity and choice

Once political decision making acquires the levels of complexity that characterise contemporary national or supranational politics, part of the reason why decision making remains possible at all is that the institutions of the political system process vast numbers of issues and decisions in parallel. The factorisation of choices, in other words, allows policy making organisations to take decisions simultaneously. Governments deal with energy crises, health care reform and urban crime at the same time. In this context, the organisation of political institutions such as administrative departments or legislative committees plays a pivotal role. They shape the perceptions and task environments of the policy makers and thereby channel and reduce the amount of processed information. But as the following discussion will show, policy issues also interact with the legislative and administrative organisation of policy making institutions in more complex way.

Reducing complexity

The institutionalisation of decision making substantially elevates the capacity of political systems to identify and process problems and solutions. One of the most central effects of organisations in politics is that they enable the parallel processing of a huge amount of information and decisions and thus overcomes the limitations of serial processing, or the 'bottleneck of attention', to borrow Simon's (1985: 302, see above) term. As Allison (1969: 698) notes, 'government perceives problems through organizational sensors. Government defines alternatives and estimates consequences as organizations process information.' In this process, 'institutions often 'solve' what Simon termed the design problem', Jones (1994b: 159) notes, 'they structure situations so as to limit choices to a relatively small number of alternatives, usually doing so by causing participants in the institutions to focus on a limited number of evaluative dimensions'. These biases are sustained because organisational units of the political system specialise in obtaining and communicating information that fits their existing perceptions and legitimises their tasks. They reduce uncertainty by forming 'negotiated environments' (Allison 1969: 701) and limit the range of considered choices to the 'recombination of a repertory of programs' (March and Simon 1958: 150). The perceptions and actions of organisational actors are thus substantially shaped by the context in which decisions are taken. 'The organisation', Jones argues (2001: 131) 'becomes our relevant referent, in effect selecting the attributes that order our decision making'.

Factorisation of choices thus allows policy making organisations to take decisions simultaneously and in quasi-independence. But this ability to respond through the decentralisation of decision making also means that the total system of decisions is factored 'into relatively independent subsystems, each one of which can be designed with only minimal concern for its interactions with the others' (Simon 1973: 270). The departmentalisation of choice in organisations means that the focus of attention is 'a function of the differentiation of subgoals and the persistence of subgoals' (March and Simon 1958: 152). Since 'most information relevant to top-level and long-run organizational decisions typically originates outside the organization, hence in forms and quantities that are beyond its control', Simon (1973: 271) argues, coherence of political choices across organisational domains is difficult to attain. As a result, multi-dimensional policy issues are rarely dealt with in terms of all their potentially conflicting facets, and choices over these policy issues rarely reflect all the potentially conflicting interests to which the issues may give rise. Instead, policy making organisations, such as specialised regulatory committees or functionally organised administrative services, regularly assure that only one or few aspects of an issue are taken into account at any point in time and that choices reflect only a limited set of interests. Under these conditions, strategic coordination of policy across the organisational structures of government often becomes very difficult. 'If organizations are permitted to do anything', Allison (1969: 700) remarks, 'a large part of what they do will be determined within the organization'.

Policy venues

The institutional channels through which political issues are processed are therefore decisive in determining how political issues are decided. They include points of access to policy agendas. But more importantly, policy venues encompass the political institutions that are formally assigned jurisdiction over policy choices. As noted above, each policy venue can be analysed in terms of its 'decisional bias, because both participants and decision making routines differ' (Baumgartner and Jones 1991: 1047) from one policy venue to the next. In legislative politics, one particular organisational effect in political decision making is the 'structure induced equilibrium' (Shepsle 1979; Riker 1982). Instead of comparing and evaluating policy choices across multiple dimensions, the organisational basis of politics guarantees that preferences are evoked and ordered by virtue of legislative committee assignment. This way, preference formation generally falls in line with a specific perception of the issues, which in turn depends on the organisational delineation of policy jurisdictions. In sum, Jones and Baumgartner (2002: 299) propose that 'committees often represent, in gross terms, different approaches or perspectives toward the issue: they are institutionalised frames'. The foremost function of policy venues is thus compartmentalisation.

Policy venues create stability for most issues most of the time because political decisions are typically taken in relative independence from each other and without reaching an agenda status that calls for the comprehensive reconsideration and coordination of conflicting interests. But when issues rise high on the agenda of political organisations, Simon (1973: 270–271) notes, 'parallel processing capacities become less easy to provide without demanding the coordination function that is a primary responsibility of these levels'. When the upper echelons across a political system address a policy issue, the parallel processing of interests and ideas is interrupted and responsibilities are often rearranged as the result of a more comprehensive consideration of the issue at stake. While the literatures on policy venues and structure induced equilibria primarily emphasises how the organisational foundation of policy making contributes to more predictable types of interest representation, the reverse logic therefore applies as well. When issues are reassigned, the balance of power shifts dramatically and policy choices often change in response. In short, the change from one policy venue to another can have transformational effects.

From this analysis of political institutions, several implications for a framing perspective follow. Not only can we expect that policy choices typically reflect the organisational biases of the administrative and legislative units of the political system involved in the decision making process. The parallel processing of decisions in politics also reinforces policy frames once they are adopted. Conflicting perceptions that would give rise to diverse interests are typically organised out of the political process. From a framing perspective, 'political institutions serve to sustain attention to particular goals over extended periods of time. In essence, they fix attention on a limited number of aspects of a situation, thereby defining and structuring issues. They do so both by factorising complex decisions and by disempowering coalitions... that would like to change the status quo' (Jones 1994b: 164). As a result, policy frames can become partially self-reinforcing and more likely to remain steady even under growing pressure from the totality of affected interests. Turning this logic around, when political actors or coalitions contest a prevailing policy frame, the effects of their advocacy will be infinitely stronger in conjunction with a shift in policy venues (Baumgartner and Jones 1991). Framing strategies are hence significantly more likely to succeed if the contesting coalition not only manages to challenge the prevailing framing of an issue in public or political discourse. In addition, it is often crucially important that frame contestation also affects the choice of policy venues in which the issue is being processed. Opportunities for venue change typically arise when issue are high on the political agenda. Once policy issue become publicly contested, administrative or legislative actors and institutions often start vying over jurisdictions and add an organisational dimension to the mere contestation of problem perception. Some effects of policy framing, however, reach beyond the organisational structure of politics. 'Therefore', Peters (2005: 355) points out, 'as we begin to conceptualize the numerous factors that might be utilized to define problems, we need to think about a broad range of variables, rather than confining our attention to those familiar labels of policy areas and government departments'. While both play a role, as the above discussion has set out in some detail, policy framing theory considers the contours of policy fields and the institutional areas that contain them as variables of the policy process, which are themselves subject to pressures and change. The next section first addresses ways in which policy framing can affect the political interest representation in a given field of policy. The subsequent section finally turns to dynamics of conflict and consolidation on the systemic level of policy making.

Collectiveness and coercion

The previous paragraphs have outlined how organisationally entrenched policy frames reduce problem complexity by several orders of magnitude, define actors' respective stakes in the issue and demarcate a decision's scope and applicability. As a result, policy frames not only structure the way policy making systems process information. By identifying the stakes in policy issues, policy frames also help to define or redefine groups and can transform their ability to mobilise interests effectively. Both in the short term and over time, policy framing thus shapes the politics of policy adoption. This section brings the interventionist and coercive aspect of public policy into focus and asks questions of how policy framing impinges on the organisation and representation of interests in the policy process by conferring incentives and resources for collective action. By analysing policy frames in terms of their structural effects, the discussion further contributes to our understanding of how the initial framing of an issue can create partially self-reinforcing dynamics on the level of political interest organisation.

Policy types

Long before the place of political science in policy analysis turned into an academic debate, Schattschneider (1935: 288) formulated one of the key propositions of policy analysis in a study on US tariffs by concluding that policies create politics. Taking his lead from Schattschneider, Heclo (1972: 104) argued in a classic review that political science 'has shied away from examining the effects of public policy' on decision structures and processes. Policy effects on 'the political system itself', he finds, raise fundamental 'problems of political rationality'. Despite frequent references to it, the general argument Heclo made has failed to transform the predominant perspective on policy making in political science. Policies remained commonly defined as the (dependent) outcome of other (independent) political factors, as substantive or material choices distinct from the politics that creates them. A corollary to this approach is that most policy analysis does not derive from any analytical concept of policy or policy issues. Lacking such a concept, the study of policy effects is severely hampered. By way of addressing mechanisms that causally link the levels of policy and politics, Lowi (1970: 318) criticised the dominant 'permissive, all-inclusive, nondefinitional approach to policy', which provides the researcher with nothing distinct in the phenomenon to study and thus, in his analysis, inevitably produces unsystematic policy studies of the 'politics of this-or-that-policy' (Lowi 1964: 715). His research turned to typologies as a tool for seeking out the basis of policy classification and tried to reveal 'the hidden meanings and significance of the phenomenon, suggesting what the important hypothesis ought to be concerned with' (Lowi 1972: 299). From this perspective, famously paraphrasing Schattschneider, Lowi (1972: 299) systematically elaborated and promoted the argument that 'policies determine politics'. By way of choosing a point of theoretical departure, Lowi (1972: 299) argues that the patterns of policy making reflect two dimensions of 'the most significant political fact about government': coercion. From the perspective of social and political actors, he reasons, the likelihood of coercion can be remote or immediate. The further a policy moves along this dimension from remote to immediate coercion, conflicts are less likely to be contained by political parties and the general levels of electoral organisation. The result is instead the direct political mobilisation of those affected. Yet this mobilisation takes distinctive forms depending on whether the mechanism of coercion is particularised (distributive and regulative policies), or applied though the reconfiguration of the environment of conduct (constituent and redistributive policies). The more particularised the coercion, the more the conflicts are likely to cut across ideologically shaped and centralised forms of political and interest representation and cause actors and interests to disaggregate and decentralise as a result.

Several aspects of this branch of policy research and the discussion it initiated are highly instructive for a framing perspective. Much of the criticism Lowi's typology provoked argues that his policy types are too rigid and that the analytical categories in reality overlap (e.g. Spitzer 1987, 1989). This criticism, however, partially misreads both Lowi's formulation of the typology and, even more evidently, his intent. In fact, Lowi appears not only to have been aware of, but also rather intrigued by the fact that most cases of policy exhibit both non-exclusive typological characteristics, and that issues sometimes move across the typological cells of his model during the political process. One of the main uses Lowi (1964) makes of the typology in his original article is to show how the policy issues are redefined in the political process in ways that correspond to shifts in the typological scheme he created. The primary purpose of Lowi's analytical distinction is thus not to generate fixed and empirically accurate categorisations. Criticism of overlapping policy types and, more specifically, the frequent claim that even redistribute policies regularly entail costs and thus necessarily include elements of redistribution, therefore partially miss the point. As Majone (2003: 311) clarifies, in such cases the key question is whether the regulatory or the redistributive aspect of the policy counts as the main policy objective and dominates the debate. Whichever aspect of the policy becomes the focus of attention will most likely also affect the resulting organisation of political interests. Criticising approaches that attempt to account for policy choice without taking into consideration how the perception of policy types is a strategic asset in the legislative process, Lowi (1964: 688) thereby links the argument back to the definition of policy issues as the causal factor. His main argument is that political interaction is structured by the expectations that actors form given the policy issue at stake. Concluding one of his case studies, Lowi (1964: 682, original emphasis; see also 1964: 695) reiterates, 'The outcome depended not upon compromise between the two sides in Congress but upon whose *definition of the situation* prevailed.'

Lowi clearly acknowledges that a policy can have characteristics of more than one type, the difference often being one of degree (1964: 713), and that types are not static. But this is not where policy typologies break down. This is precisely where they produce unique explanatory insight of relevance for framing analysis. In two articles, Kellow (1988, 1989) elaborates how focusing on the role of issue definition allows the researcher to account for political organisation and interaction as the result of political agency. 'Lowi's analysis is important in focusing attention on the ways in which policy contents structure political interaction', Kellow finds (1989: 541), 'but I am sure he never meant us to conclude that political factors were irrelevant to the ways in which policy proposals were structured'. Following Schattschneider's dictum that issue definitions constitute the choice of political battlefield, Kellow (1988: 715) instead argues that policy dynamics must be accounted for as the 'result of political manoeuvring over the definition of the issues'. Policy typologies hence stop well short of giving a full explanation of policy making, but greatly contribute to 'a view which accounts for the dynamism of the policy process and which provides a role for politics in shaping policy proposals, which in turn influence the political struggles over policy adoption, which influence policy outputs' (1988: 721). The most effective use of his scheme, Lowi proposes accordingly, is by 'looking through the eyes of the top-most officials at the political system and how and to what extent the system shifts obliquely as their view of it shifts from one policy prism to another' (1972: 300).

In similar ways as the literature on policy venues discussed above focuses the analysis on the effects of issue redefinition on the institutional reallocation of responsibilities, Lowi's work can be read as a way to conceive of policy dynamics rooted in shifts of interest mobilisation and political competition that result from the same effects. Along these lines, later advances in policy research pushed the same point even further. Though acknowledging the importance of Lowi's analytical categories, Wilson (1973, 1980) for instance criticises Lowi's formulation for not having moved completely beyond substantive policy types, and calls the analytical scheme 'ambiguous and incomplete' (1973: 328). Echoing some of the prevailing criticism, he points out that it is difficult to fit many policies into the cells, while other substantive policy areas deviate markedly from the predicted political process. In a variation of Lowi's typology, he proposes to redefine policies along the dimension of costs and benefits and differentiates between their respective diffusion and concentration. Each combination of characteristics leads Wilson to expect a specific type of politics. When the costs and benefits of a proposed policy are widely distributed, interest groups have no incentive and little power to advance their particular causes: majoritarian politics prevails. In the opposite case, when costs and benefits are concentrated, the general public is less involved and more directly affected interests are likely to organise and seek political influence. Interest group politics is transformed into client politics when costs are diffuse, in which case interest groups have high incentives to influence policies from which they profit without creating an easily identifiable bearer of the costs. Politically organised resistance, Wilson (1980: 369) argues, is unlikely to form in this case. The rapid proliferation of public interest organisations in some sectors, however, can transform client politics back into quasi-interest group politics. Lastly, when costs are concentrated and benefits are diffuse, political mobilisation is unlikely to happen to begin with. Political entrepreneurs can take up such issues, Wilson concludes, mobilise latent interests and define the cause in such a way as to counteract the readily organised resistance of the bearers of the costs.

Creating policy interests

In important ways, policy typologies do not make it possible to deduce straightforward conclusions for individual cases of policy making, but they conceptualise aspects of policy design with which framing analysis should be concerned. Wilson's argument, for example, stops far short of a deterministic interpretation of the effects of the distribution of costs and benefits on the resulting nature of the political organisation. The cell occupied by entrepreneurial policy in his typology would otherwise have to be empty. Instead, as Wilson (1973: 335) points out, major legislation of this type is passed after 'the successful mobilisation of a new, usually temporary, political constituency'. Such mobilisation occurs despite the mechanisms identified in his scheme, not because of them.

Based on his analysis, Wilson (1973: 337–338) also notes that the fact that 'organized groups may be more active in political institutions where power is diffuse does not necessarily mean that they will be more effective in such places'. Lacking a coercive function such as through a strong party system will open up multiple points of access across the institutions of government, but in a non-discriminatory way. The chances for some interests to prevail over others and co-opt political institutions must therefore depend on other factors. 'Furthermore', Wilson notes, 'in a political system where power is fragmented, individual officeholders are freer to choose which organisation to heed, if any'. In such cases, his analytical scheme enables the identification of political problems that must be overcome in order to counter the effects of concentrated costs and diffuse benefits. Similarly, Wilson (1973: 341) finds that the strong organisation and mobilisation of interest politics in cases of both concentrated costs and benefits does not prevent policy makers from pursuing certain goals. While the underlying mechanisms that Wilson identifies are general, the specific costs and benefits that structure political organisation in this analysis are in no way pre-political or in any other way given. Instead, they reflect the conflicts of interests that follow from a specific policy proposal (1980: 371–372), and it is through the redefinitions of policy issues (1973: 330; see also 1980: 384) that policy makers affect the process of political organisation.

Some political interests are hence only identifiable in relation to their common stake in a policy and only form in response to initial policy making in a certain area (Pierson 1993: 598, Wilson 1980: 334). Other groups exist prior to a policy initiative, and in such cases their internal organisation and the politically created distribution of costs and benefits becomes critical. Even very slight changes in the policy approach, such as changing the basis for the distribution of benefits, or changing the regulatory approach, can render it difficult for existing interest organisations (and especially economic interest organisations) to identify their common goals. Changes in policy design may split members' interests outright. When that happens, it may be too difficult or too costly to seek new forms of interest representation and respond to the displacement of conflict. Alternatively, the members of organised groups may simply be too slow to assess the full consequences of newly emerging lines of policy contestation to be able to voice their position and use their resources effectively. The lack of appreciation for such structural effects of policy formulation on interest formation leads Pierson (1993: 627) to identify a 'need for a more fundamental reexamination of research agendas' to gain more analytical leverage. Using a long-term perspective, Pierson's (1993: 598) analysis points out that 'interest groups often seem to follow rather than precede the adoption of public policy'. These findings leave him to question the direction of standard causal arguments in policy analysis. Instead of conferring benefits on previously organised groups who have presumably pressured government institutions for recognition of their claims, Pierson finds the logic reversed. It is the policies themselves that create the benefits and thus provide incentives that cause groups to identify themselves as beneficiaries and seek more organised forms of representation in response. From this perspective, 'major policies can be seen as rules of the game, influencing the allocation of economic and political recourses, modifying the costs and benefits associated with alternative political strategies, and consequentially altering ensuing political developments' (Pierson 1993: 596). While established policies have structural effects on political agency and organisation, as Pierson (1993: 596, 608) argues, similar effects can result from policy initiatives in the short term. Both perspectives are not mutually exclusive, and a full analysis of the effects of the framing of policy choices must consider both levels. Policy framing that affects the distribution of costs and benefits or impinges on the mobilisation and structuring of interests can hence cause policy issues to develop new momentum. By generating affirmative feedback or disabling adverse alliances, the framing of policy issues can partially create the political environments in which the issues are processed. Strategic policy makers, in other words, will attempt to frame policy initiatives so as to trigger the political dynamics to which they will subsequently be subjected, and pursue their interests by creating political dynamics that facilitate them.

In the case of EU, these dynamics are of particular concern since a significant number of the legislative initiatives create policies where none existed so far. Under such conditions, policy framing plays a crucial role because it affects the original formation of political interests in the respective policy fields at the supranational level. These short-term effects of policy framing in the EU thus closely resemble long-term, structural policy feedback in more mature policy making systems. The framing of the issues is central to the analysis of EU policy making in that it settles the scope and nature of the applicability of new rules and regulations, and in doing so defines the new or revised policy areas and the interests to which they will give rise. Furthermore, its bias for regulatory policies (Majone 1994: 85–92) means that EU policy making predominantly gravitates towards the policy type which Lowi identifies as having the strongest tendency to disaggregate and de-centralise the political organisation of interests (Traxler and Schmitter 1995). Regulatory policies undermine the influence of party ideologies and territorial political representation, such as by member state governments, and thus leave the supranational actors with more leverage in shaping political interaction constellations. How policy issues are structured, and how their framing evokes certain policy dynamics rather than others, can thus be understood to precede rather than follow the organisation and alignment of actors and interests and direct or divert the ways the vectors of political power are pushing.

Conflict and consolidation

One key question raised at the outset of the discussion of policy dynamics in framing analysis concerned the problem of how frames can exert stable and systematic effects throughout the entire policy process. The arguments outlined so far have highlighted that the organisational level of politics is critically important in determining how frames are generated, and how framing dynamics are structured and sustained. The discussion of analytical concepts of policy has furthermore emphasised that the formation and representation of interests in a given polity often reflects, or responds to, which facets of the policy design are emphasised in public discourse at any given time. Policy framing that addresses these aspects of policy design and political organisation were identified to be particularly effective in the policy process and are, at times, expected to give rise to lasting restructuring of policy constituencies. While these dynamics characterise the interplay between policy frames and structural factors at various levels of the political system, policy framing can easily remain contested. No matter how entrenched in the political institutions and the structure of organised interests, any framing of a policy issue can be challenged. Building on the previous discussion of complexity, choice, and the political construction of interests, the next section discusses system-level framing contests that occur when contentious issues rise to the top of the political agenda.

Serial policy shift

Following the logic of parallel processing in political systems summarised above, the choices over most policy issues are made in relative isolation from one another. As long as systems of limited participation prevail, the framing of the issues rarely changes. Specialised venues of policy making and organised interests instead converge around a common identification of the dominant evaluative dimensions in a given area of policy. The resulting choices often represent local policy equilibria that reflect limited attention to only selective attributes of the issues and the interests with which they correspond. Such local equilibria can veer far from policy positions that could facilitate majority support from across the entire political system under conditions of wider participation and re-evaluation of the criteria for policy choice. In the case of the EU, however, the reviewed literature on policy communities has revealed that stable coalitions of interests, and the type of institutionalised cooperation between political actors and interest representation that are characteristic of the arrangements that maintain local equilibria, are rare. Given the fluidity of interest representation in the EU, systemic policy dynamics appear to play an even more frequent role in policy decision making than in most national political systems. Consequentially, it is critical to understand how framing dynamics can shift when issues are pushed up the decision making hierarchy.

When policy issues rise on the political agenda, Simon (1973: 270–271) notes, 'the bottleneck becomes narrower and narrower as we move to the tops of organisations, where parallel processing capacities become less easy to provide without demanding the coordination function that is a primary responsibility of these levels' (see also March and Simon 1958: 150). Here, parallel decision making shifts to sequential, or serial, processing of information as issues are addressed outside their original decisional context. While 'particular decision domains

will evoke particular values', Simon warns (1983: 18), 'great inconsistencies in choice may result from fluctuating attention'. New
salient dimensions of choice can override the dominant framing of
the issues, and times of stable policy choices give way to dramatic
change as a result. Jones (1994b: 184–186) refers to this type of change
as the 'serial shift' in policy making to emphasise the change in the
information processing from parallel, compartmentalised venues to a
comprehensive reconsideration of policy issues. Organisation theory
explains why shifts from one institutional venue to another can bring
about considerable changes in the ways policy issues are framed and
decided. Yet, Baumgartner and Jones (1991) point out that the reverse
logic applies as well. When policy issues are re-evaluated and the focus
shifts from one attribute or dimension to another, conflicts cannot
be easily contained within the original decision domains. Political
systems often react to such inconsistencies by reallocating issue responsibilities or by reorganising or creating new legislative or administrative venues that more adequately reflect the new salient problem
perception. While this type of policy shift will often appear erratic,
some theoretically derived determinants of greater or lesser political
volatility can nonetheless be identified. The following discussion
reverts back to Schattschneider's logic of conflict displacement and
asks how framing analysis can address questions that concern the
consolidation of political forces at the systemic level of policy making.

Conflict of conflicts

Despite Schattschneider's interest in politics as a volatile process of
contestation, nothing in his analysis suggests that the business
of governing is futile. Instead, it is the deliberate management of
the lines of conflict, their 'exploitation, use, and suppression' (1957:
935, 1960: 65), which allows policy makers to shape the political
process and its results. 'We are concerned here', Schattschneider
(1957: 935) notes, 'with the use of conflict to govern, the use of conflict as an instrument of change.' Every dominant conflict, Schattschneider explains, not only divides, but also unites diverse interests
and actors as they identify common stakes in the conflict and ally
in pursuit of their shared objectives (1957: 940, 1960: 62). In this
framework of analysis, political consolidation is therefore the flip
side – rather than the opposite – of political conflict (see also Mair
1997). Yet how this process plays out, and with what effects on the

policy process, varies considerably. Depending on the structure of the issue at stake, political consolidation can entail marginal adjustments or take the form of political restructuring. In the case of policy issues that encompass greater numbers of unrelated underlying dimensions, serial policy shifts are more likely and their effects are likely to be more dramatic (see Baumgartner and Jones 2002: 15). When the attention shifts from one evaluative attribute to another, entirely unrelated attribute, the reframing of the issue recasts the conflict in such a way that the existing coalitions easily lose their footing. As a result, the higher the number of unrelated evaluative dimensions that characterise an issue, the more easily the prevailing issue representation can be challenged. Most importantly, however, the framing perspective places special emphasis on the role of politicisation. Just as unrelated issue dimensions facilitate serial shifts, one coalition can seek to displace the entire conflict, thereby silencing their adversaries and subordinating one conflict to another. This framing strategy, Schattschneider (1957: 939) explains, equals 'a flank attack by bigger, collateral, inconsistent and irrelevant competitors for the attention and loyalty of the public'. The general underlying dynamic, the 'conflict of conflicts' in Schattschneider's (1960: 60) terminology, means that in every political system an entire universe of alternative, inconsistent conflicts and interests is subordinated and finds little or no expression in the political process. 'All depends on what they want most', Schattschneider (1957: 940, 1960: 66) summarises, 'what they want more becomes the enemy of what they want less'.

Issue framing never take place in a political vacuum. Some dimensions of conflict are more established or more easily evoked than others. Different systems of political representation favour different ways of consolidating diverse ideas and perceptions over others. But political systems facilitate or predetermine to widely varying extents which system-level conflict space has sufficient force to recast issue framing controversies. In the case of the EU, lacking the clear dimensionality of a stable political space, but with multiple constituencies and a complex system of representation, issue politicisation provides particularly rich opportunities for serial policy shifts. 'Populist policy equilibria' (see Jones 1994b: 158, 174–175) that reflect mass preferences over policy issues are notoriously difficult to seek out. Competing supranational actors therefore often have reason to further expand and politicise framing conflicts that allow them to aggregate increasing numbers of interests. Under these conditions, the framing perspective indicates that policy politicisation can prevail at the supranational level with particular vigour. Expanding the scope of a policy controversy may be the only mechanism to create a stable majority. As noted above, Schattschneider's primary interest in the strategic expansion of the scope of a conflict concerns not only its mere magnitude, but the fact that the lines of conflict shift as the scope of a policy controversy changes to incorporate new actors and interests. Frame contestation that mobilises diametrically opposed interests can sometimes consolidate a conflict, instead of opening up opportunity for policy change. More typically, however, persistent politicisation of an issue over the long run will, in most democratic and pluralist political systems, eventually causes one or both camps to expand the scope of the conflict and thereby shift the direction of the conflict as well. Depending which way the political ground shifts, it normally favours one side in a political controversy and sometimes creates a new majority position. With a single predominant conflict of conflicts largely missing in the EU, the discussed weakness of the dimensionality of the EU political space means that the recasting of policy conflicts by way of issue framing may well be much easier and more consequential than in traditional political systems. Whether the framing and reframing of the issues at stake results in recurring serial policy shifts, or the consolidation of conflicting positions, or stalemate, will then depend largely on the characteristics of the policy issues and the framing strategies of the actors involved.

Summary

The preceding paragraphs have introduced Schattschneider's (1957, 1960) view of politics as the management of conflict. According to this perspective, policy makers can affect the ways in which interests turn into conflicts, cleavages, majorities and political opportunities for change by defining what is at stake and for whom. Reframing policies can thus cause 'oblique turns in politics' (Lowi 1972: 300). Alliances can form, rise in importance, or crumble as a result of how policy issues are framed and reframed. Models of bounded rationality (Simon 1985, 1987, 1995; Jones 1994b) and the heresthetic manipulation of decision making (Riker 1986, 1990, 1995) have greatly advanced our understanding of why the complexity of political choice and the limits of rationality render framing critical for the outcome of political decision making. Yet while frames structure choice, frames are not given in a politically meaningfully way. Instead, frames are contestable in most political contexts. Conflicting interests remain partially ordered and imperfectly reconciled. There is room and incentive for manipulation. As a result, the scope and nature of framing effects in the policy process requires separate explanations.

Sometimes external events cause the overall perceptions of policy issues to change dramatically and result in the reshuffling of policy positions. But politics entails more subtle and frequent forms of framing and reframing, which develop their impact endogenously in the process of policy making. The above discussion has shown how the organisational foundation of politics can structure and sustain emerging policy frames. Framing strategies that reorder jurisdictional responsibilities and translate into the empowerment of specific policy venues thus limit the possibilities of frame contestation and can help to keep framing effects steady, even under increasing pressure for change. Framing further exerts structural effects if policy frames translate into mechanisms that confer resources and incentives, or otherwise impinge on the formation and representation of political interests. Because policy choices involve varying degrees of coercion, the mobilisation of pressure and the elevation of certain lines of conflict over others do not necessarily hinge on overtly partisan agency. 'Exactly who is induced to mobilise', Pierson (1993: 599) stresses, 'will often depend on the precise nature of the policy intervention'. Policy makers are in a unique position to assess and anticipate the political effects that changes of issue representations have on individual interests and the wider policy field. The right to formulate policy initiatives thus gives political actors a critical advantage over organised interests and allows them, within limits, to strengthen or weaken the interests with whom they subsequently engage in policy deliberation. Policy framing that manipulates the scope of an issue has been emphasised as a particularly potent mechanism to cause the lines of policy contestation to shift. By defining the scope and applicability of a policy, issue framing can add or subtract actors and interests, and thereby construct the areas of likely agreement as well as the areas of conflict that shape coalition formation processes (see also Sebenius 1983).

While the theoretical exploration of framing effects in policy research presented in this chapter draws on important contributions from a wide variety of relevant subfields in political analysis, it can hardly claim to be exhaustive. As framing research continues to explore ways to connect the structure and representation of policy issues to ensuing policy dynamics, this field of research will develop beyond its already burgeoning diversity to emphasise new and distinct perspectives. To conceptualise framing effects according to their functions in the policy process and ask how they affect the role of the organisational foundation of politics in the reduction of policy complexity, the creation and organisation of collective interests, and the structure of political conflict and competition may serve as a way to organise more focused compactions of the different levels of analysis on which policy framing not only develops its own momentum, but also interacts and competes with factors characteristic of each political system.

3
Biotechnology: The Creation of a Policy Field

Overview

The field of biotechnology

Modern biotechnology refers to the techniques of genetic manipulation, modification and recombination, the processes collectively described as genetic engineering. The primary function of these advances in biotechnology is the transfer of genetic information or the altering of genetic material in ways that do not occur in nature. Today, the areas of research and development in which biotechnology can be employed are so various, and constantly expanding, that it has become customary to group them together and distinguish them roughly by the different types of end products. According to this categorisation and terminology, 'red biotechnology' refers to the application of biotechnology for medical purposes, for example in the production of antibiotics and insulin. 'White biotechnology' covers the industrial use, such as in the production of decomposable plastics that are beginning to replace traditional petroleum-based products, or in the production of ordinary washing powder. Finally, 'green biotechnology' comprises the use of the technology in agriculture and includes for example the development of new crop plants.

The empirical focus of this chapter is on the legislative dynamics in the area of green, or agricultural, biotechnology in the European Union from its beginnings in the 1980s until the revision of the regulatory framework during the reforms of the 2000s. At the centre of the EU biotechnology policy making in this area, especially during the initial years, lies the regulation of genetically modified organisms and microorganisms. Organisms of the first type, often referred to by their acronym GMOs, are biological entities capable of replicating or transferring genetic material, which has been altered in a way that does not occur by mating or natural recombination. Certain variants of crop, such as maize or soybeans, are among the most common commercialised genetically modified organisms in use today. The main commercial research in green biotechnology focuses on the development of herbicide-tolerant plants and plants that are resistant to certain insect pests. In order to develop crop plants with such characteristics, particular genes are transferred from one organism to another, including between non-related species. Microorganisms, on the other hand, are living microscopic entities that can play important roles in manufacturing processes such as baking and brewing, in food processing and as food additives. Genetically modified microorganisms (or GMMs) are primarily used in the production of food to increase nutrition values, slow down the process of spoilage, or enhance flavour.

The policy issues

The recent surge of commercial interest in genetic engineering has made biotechnology one of the most rapidly advancing fields of scientific inquiry in modern times. Yet, as recent scientific research has greatly advanced human understanding of these processes, it has simultaneously posed the question of how far the ever-widening possibilities arising from biotechnological research can and should be explored and exploited. As the following discussion of EU policy will demonstrate in detail, the political and legislative responses to the rapid advance of biotechnological research are just as varying as its applications. Regulatory approaches to biotechnology policy in the EU sometimes overlap, conflict, or compete with policy in the same area that takes a different approach. Most importantly, however, far from all legislation relevant to the field of biotechnology makes explicit and exclusive reference to the technique of genetic modification itself. Instead, many policy designs focus on the final product, a certain type of food or seed for example, indiscriminate of the specific processes by which it was developed. Such policy making strategies formulate regulatory standards specific to the nature of the product (and the way in which it is potentially hazardous) rather than the nature of the biotechnological processes involved in manufacturing that product. Because technology-based regulation often cuts across different product areas, it is also referred to as *horizontal legislation*. Sector-based or *product legislation* by contrast is sometimes identified as vertical policy. Most regulatory regimes in the field of biotechnology are combinations of both policy making strategies. During the period of EU biotechnology policy making discussed here, the choice over the right mix of these different policy approaches – and the question of whether one approach should entirely replace the other – was in itself one of the key issues of political disagreement. As the empirical discussion will show, the political implications of the choice between the two policy approaches often far outweighed the regulatory implications. Following a similar logic, the substantive regulatory issues addressed in the analysis of EU biotechnology policy making are often of such a technical nature that it is difficult to convey at the outset a sense of the political conflicts that they caused. The following overview of the substantive policy questions that most frequently reoccur in the following chapters is meant to provide a very basic level of familiarity with the substantive policy issues before the political dimensions of EU biotechnology policy take centre stage.

The two earliest EU laws discussed here regulate the use of genetically modified organisms and microorganisms in research and agriculture. The applicability of the two laws depends on the degree to which the public and the environment are exposed to the biotechnological product in use. Accordingly, EU directive 90/219 addresses the use of genetically modified microorganisms for research, development and production in laboratories and industrial production under conditions that qualify as *contained use*. The main purpose of this directive is to provide safety precautions against the risk of escape of these microorganisms. In this context, the main procedures outlined in the law describe the risk assessment these products have to undergo before they can be used in research laboratories as well as the formal procedure of the authorisation of their use in research and development.

The *deliberate release* directive 90/220, to the contrary, deals with the experimental field release of genetically modified organisms, such as genetically manipulated plants, and their marketing as a product. As in the first case, the authorisation procedure for the field release or marketing of genetically modified organisms includes a specific environmental risk assessment. Both the procedural question of granting or denying approvals and the substantive question of risk assessment remained contested from the inception of EU biotechnology policy until the later revision of the entire regulatory framework. One key issue in this context concerned the question of whether products such as genetically modified food should require *labelling*. If so, what should be the appropriate scope and purpose of any such labelling scheme? What information should labels contain? Should they imply warnings, and based on what criteria? The presence of only very low levels of under 1 per cent of genetically modified crops, for example, can easily result from accidental contamination during harvest, transport and processing. It may also result from natural cross-fertilisation in areas where genetically modified plants are grown. Depending on the thresholds levels, strict labelling provisions would therefore imply the possibility of separating organic crops from genetically modified variants entirely. That may not be possible. Linked to the question of labelling is the issue of *traceability* and, ultimately, *legal liability*. Once a genetically engineered product has been approved for its deliberate release, should records be kept that allow authorities to check that labelling rules are followed? Also, if an authorised product should prove hazardous at a later stage, despite the initial assessment, traceability could facilitate focused counter-measures and, in extreme cases, a complete recall of the product. Many of these issues however, can only be addressed comprehensively if the separate components of the regulatory scheme are designed in complete anticipation of the potential problems and the resulting regulatory demands that might emerge at the various stages of a long and complex process that leads from research laboratories to supermarket shelves. As a result, the level of policy coordination necessary to address these questions with a reasonable degree of internal consistency is high.

Throughout the better part of more than two decades covered by the following legislative case studies, one specific question of EU policy making concerned the *patentability of biotechnological inventions*. Legal protection of patents is a key aspect of a functioning common market for any good. EU level harmonisation of biotechnology patents was largely initiated due to concerns about divergent national legal approaches to this question and their effects on biotechnological innovation in Europe. Over time, however, problems specific to the area of biotechnology came to play a more dominant role in the political decision making process. Questions of what should count as an invention, given that biotechnology deals with the manipulation of natural biological entities and processes, loomed large. Unsurprisingly, arguments that EU policy on the legal protection of biotechnological invention should be limited in scope were made on ethical grounds. The ambiguities of the issue at stake made the choice over biotechnology patents one of the most hard-fought political battles in EU legislative history. They also provide an ideal illustration of how questions on legal and regulatory detail could trigger political upheaval and transform policy making in the EU for years to come, well beyond the field of biotechnology. The next section provides a very brief history of the empirical developments and a preview of the main analytical points.

A short preview

In 1988, after five years of deliberation, the European Commission formally adopted the first proposals for the two EU biotechnology directives 90/219 and 90/220. In the years that followed, two more major biotechnology laws were passed, the directive on biotechnology patents and a special regulation on novel food. By 2003, each of the four policies had undergone reassessment and, for the most part, dramatic revisions. The following discussion will argue that EU legislative policy making from the mid-1980s until 2003 was characterised by three distinct policy dynamics. First, the creation of the policy field, with the adoption of the two directives 90/219 and 90/220 (Council of the European Union 1990a, 1990b), marked the success of a specific policy frame. This frame focused on human and environmental risk posed by the use of genetic engineering as a technology in research and development. Legislation under this horizontal frame set out regulatory rules largely indiscriminate of the type of biotechnological application. The adoption of this policy frame was a highly contested political choice over which Commission services clashed internally throughout the process of policy formulation. Once the regulatory scheme was established, however, it was constitutive of a new form of political interest representation at the European level. Specifically, after their initial defeat in the EU policy process, the affected biotechnology industries regrouped effectively in response to the adopted legislative framework.

In close cooperation with the Directorate-General (DG) for Industry, the newly founded lobby worked towards making biotechnology an integral part of the Commission's focus on economic competitiveness and growth, which shaped the entire policy agenda of the Commission from the middle of the 1990s onwards. Under this new frame, the policy strategy was to exempt biotechnological products from the horizontal safety directives, formulate less stringent regulatory standards and promote the sectors of agricultural and industrial biotechnology. Biotechnology policy was portrayed both in terms of its contribution to the creation of a common European market and in terms of its effect on Europe's international competitiveness. The political conflict over a largely deregulatory policy approach that focuses on economic competitiveness, on the one hand, and the original environmental safety frame, on the other hand, created lasting political coalitions in supranational politics. These coalitions involved various Commission services, but more importantly, the conflict promoted the rise of the European Parliament as a new political venue and as a decisive political actor that forged a coalition with highly organised environmental interest groups at the EU level. The institutional stakes in sustaining the original conflict kept biotechnology high on the Commission's legislative agenda throughout the 1990s. In a legislative process that had a combined span of over 150 months, two new policies (on biotechnology patents and novel foods) were adopted. Both laws were cornerstones of the new economic competitiveness frame in EU biotechnology policy.

But after years of wrenching political conflict, central elements of the new policy on novel food were revoked in the same year they came into force. By the end of decade, political campaigns against biotechnology formed across Europe and consumer confidence in most areas of biotechnological application had eroded. Even the proponents of the economic competitiveness frame conceded that deregulation of the technology and trimming down of the risk assessment and market authorisation procedures would only fuel public scepticism and hence further reduce the possibility of reaping economic benefits. Consequently, more stringent safety regulations were presented as the only hope to restore consumer confidence. As the perception of the issues of biotechnology regulation changed, the policy conflict was transformed as well. Eventually, the political leverage and involvement of the advocates of biotechnology deregulation declined, paving the way for the repeal, in 2001, of the central horizontal EU directive 90/220 on field releases and the marketing of biotechnological products. Accompanying policy measures that further expanded the scope of these safety regulations and repealed and replaced the novel food regulation of 1997 followed in 2003. In the end, the EU adopted by far the most encompassing and stringent regulatory regime in biotechnology worldwide. The legislative strategy to reframe biotechnology as an issue of economic competitiveness and to promote its application in the EU had almost entirely failed.

In reflection upon EU biotechnology policy making, one senior Commission official (Cantley 1995: 668) notes that 'on complex subjects... when the mass of information and opinion is effectively elicited, there has to be a radical condensation and filtering to summarise the debate into the drafting or amending of a legislative text... Both in the communication, and in the condensation, the opportunities for distortion, accidental or wilful, are legend.' The interest of the following empirical investigation is to show how the framing of the issues of biotechnology policy on the supranational level gave rise to distinct policy dynamics that structured the legislative policy choices over more than two decades.

Environmental safety policy

Defining the scope

Early Commission approach

The policy responsibility for questions concerning genetic engineering in the EU initially rested fully with DG Science, Research and Development. After groundbreaking scientific conferences on recombinant DNA (rDNA) research in 1973 and 1975 at Asilomar in California, DG Research promoted a distinctly non-regulatory approach to biotechnology and focused primarily on creating and funding new research programmes in this area. On regulatory issues, it adopted the opinion of the major European scientific organisations in the field, which argued that there was no policy demand since recombinant genetic research posed no specific risks. As the Liaison Committee on Recombinant DNA Research of the European Science Foundation expressed in conclusion of a 1981 report: 'Finally, the Liaison Committee reaffirms its opinion that there is no scientific justification whatsoever for new legislation specific for recombinant DNA research and moreover, it sees no justification for further extensive recombinant DNA risk assessment programmes' (quoted in Cantley 1995: 522).

By 1983 the Commission had concluded a review of the state of European biotechnology and adopted a Communication to the Council on the question of 'Biotechnology: The Community's Role' (European Commission 1983a). According to the analysis of the Commission paper, agricultural and industrial applications of biotechnology were progressing fast and held great promise. Europe, however, was described as lagging behind, while the US and Japan appeared much better positioned. The Community's role in a biotechnology policy framework was hence foreseen to lie in the 'promotion of agricultural and industrial competitiveness, the harmonisation of markets and the removal of barriers' for which 'a Community framework will provide the necessary security and stability'. Biotechnology was supported particularly emphatically because it was seen as 'necessary to the development of European agriculture. It opens up the prospects for the upgrading of agricultural products and to some degree the reduction of support for these products from the Community budget.' Fostering research and development on a European level was identified as a cornerstone of the proposed approach, but the Commission put the greatest emphasis on the need for the removal of economic barriers within the EU to ensure 'the prospect of a sizable internal market'. Consistent with the almost exclusive focus on the question of how to promote European biotechnology industries, regulatory issues were barely taken up at all.

The tone of a second, far more extensive Commission communication (European Commission 1983b) from October of the same year on 'Biotechnology in the Community' largely echoed the general sentiment that in terms of risk management, monitoring the research and development in biotechnology was a necessary, but also sufficient, reaction to the scientific advances in the field. Again, agricultural biotechnology was highlighted as central to the European strategy. While the authors of the Commission communication identified divided public and parliamentary perceptions concerning the trade-off between the possible risks posed by applications based on recombinant DNA research versus the promises of biotechnology, the Commission also found 'that the consumer is willing to accept innovation'. Accordingly, the foremost role of European institutions and national governments was said to lay in 'encouraging innovation, harmonizing regulatory regimes to create a genuine common market, and ensuring that regulations are based on rational assessment and well-informed debate'. In the text of the communication, the issue of risk and safety is mentioned last. 'From a first review of the situation', the Commission expressed in the closing paragraphs of the section, 'it would appear that the application of current Community regulations in the various fields [pharmaceuticals, veterinarian medicines, chemical substances, food additives, bioprotein feedstuffs] will meet current regulatory needs... On the basis of its experience deriving from the use of these various instruments, the Commission will put forward general or specific proposals appropriate to create a regulatory framework suitable for the development of the activities of the bioindustries and for the free circulation of goods produced by biotechnology.' Reflecting upon the multi-dimensional issue at hand, the communication stressed the need to involve different DGs in the policy processes. But despite the acknowledgement of divergent and possibly conflicting interests in the policy field, the Commission only sought 'to monitor the situation, and hence to concert possible policy discussions and initiatives across the services, with Member States, and other relevant groups', as proposed in the initial communication. However, shortly thereafter the policy debate changed.

Turf wars and coordination failure

The challenge the European Commission faced in 1983 was to adapt organisationally and procedurally to the complexity of the issues. Wide-ranging and interrelated features of the new policy field demanded high levels of internal coordination between the sectorally assigned responsibilities of the various DGs inside the Commission. The result was a succession and sometimes a coexistence of coordination committees and subcommittees with often contradictory agendas.

The first of them was the Biotechnology Steering Committee (BSC) created in 1984 and chaired by DG Research. The committee was designed as a forum for discussion that was open to all interested Commission services. The Concertation Unit for Biotechnology in Europe (CUBE) served as its secretariat, drawing on existing expertise inside DG Research and by reemploying staff from previous, related programmes to monitor the developments in biotechnology and to raise the awareness of those issues across the Commission. CUBE's focus on research and development was further cemented when it was put in charge of the Biotechnology Action Programme adopted in 1985. But with no decision making mandate delegated to it, interest in the Steering Committee evaporated quickly among senior officials from other DGs and the meetings of the Committee became less and less frequent. One year into its existence, the Biotechnology Steering Committee had proven to be ineffective in providing policy making impetus. DG Environment started to attend its meetings and the committee was extended to include the Biotechnology Regulations Inter-Service Committee (BRIC). Officially, BRIC served as a technical unit intended to provide further expertise and was given even fewer independent competences. As an internal memo from that time clarified, 'the individual services will retain their executive function in the examination, initiation and management of regulations as well as in pursuing harmonisation of testing regulations and information exchange with international bodies' (European Commission 1985a). In reality, however, BRIC quickly developed into the principal political body. In the first year after its creation, the new subcommittee met over a dozen times, making it virtually impossible for the Biotechnology Steering Committee as the more principal political body even to catch up with the evolving discussion. Moreover, it was obvious from the outset that the work of the sub-committee was geared towards an entirely different approach to biotechnology than the one pursued by DG Research and the Biotechnology Steering Committee. The Commission paper (European Commission 1985a) outlining the function of BRIC highlighted in its first sentence the 'European Community's interest in controlling the possible risks from biotechnology' and moved on to point out that 'it is better to evaluate potential risks as far as possible in advance' as well as through continues monitoring. This way, the communication concluded, health and environmental protection are guaranteed and industry can benefit in 'a context of increasing regulatory certainty'.

In contrast to the Steering Committee dominated by DG Research, the chairmanship of the Biotechnology Regulations Inter-Service Committee alternated between DG Industry and DG Environment. While the mission of BRIC was closely aligned with the regulatory outlook of DG Environment, the co-chairing DG Industry did not share the same policy outlook. Instead, Industry Commissioner Karl-Heinz Narjes attempted to reach out to the biotechnology industries and sent barely concealed messages pleading with them to enter the fray more forcefully. In the conclusion of a speech on industrial biotechnology in Europe, prepared in cooperation with DG Research and CUBE in 1985, his speaking notes cautioned that the creation of a regulatory regime based on the technology instead of on the product and product safety would lead to 'irrational constrains which would inhibit innovation'. Instead, he reiterated his view that, in terms of regulation, 'biotechnology does not require major changes, and that it is important not to stifle the new technology'. 'However', the Industry Commissioner warned, 'there may be instances where it is difficult to resist political pressure to do more' (European Commission 1985b).

Initially, however, DG Industry's assessment that there was little need for regulatory policy on the European level remained widely shared, as became evident during two high level consultation meetings of the Commission with national representatives in April 1986. After the meeting, a Commission paper was circulated in which it was emphasised that the 'absence of a rational basis for treating the regulation of rDNA organisms separately from the regulation of other organisms' had been emphasised during the meetings. In light of the notification procedures already in place at the Community level, the paper went on, 'many states felt that existing regulation was adequate'. Instead, research should be fostered. A joint position paper provided by five industry associations upon the request of the European Commission stated even more bluntly that it was 'neither necessary nor desirable to formulate a single set of guidelines and rules which cover every aspect of biotechnology' (ECRAB 1986: I). More specifically, industry representatives warned, 'guidelines and rules for determining the safety of products depend on their final use and must be specific to each industrial sector... It is the industry view that any additional requirements for product specification can be accommodated in the present legislation and that there is no basis for any discrimination against biotechnological products' (ECRAB 1986: 5).

In the meantime, the European Parliament had picked up the issue and became a strong and independent advocate of a more encompassing approach to tackle the prospects and the risks involved in the newly established policy field. After public hearings in 1985 (European Parliament 1985), the critical Viehoff Report, named after the rapporteur and member of the Environment Committee who initiated it, stirred the activism of several parliamentary committees. It created a diverse institutional commitment to the issue and led to a resolution (European Parliament 1986) emphasising the need for an integrated policy response. In terms of what specific direction such an approach should follow however, the Parliament's resolution was vague and lacked the robust regulatory commitment that would characterise Parliament positions in the years to come. Instead, under the first heading, the resolution cites economic competitiveness as the main objective and proposes that Community funding be made available to foster basic research with a view to exploring possible areas of application. Only under heading fifteen the resolution even acknowledged the existence of the Biotechnology Regulation Inter-service Committee and its work on risk assessment and other safety issues. Like concerns for Europe's economic competitiveness, ethical issues such as the prohibition of the manipulation of human genes figured more prominently in the policy statement than questions of human and environmental protection.

Identifying the stakes

By 1986, the issue of biotechnology had been placed on the EU agenda, mostly due to Commission activism, across the political institutions of the Community and among the affected industry sectors. Both the scope and the direction of EU policy in this area had been ambiguous and controversial from the outset. Yet, despite intense consultation and feverish attempts by the Commission to internally adapt to the various demands on Community level, it long remained unclear who was in charge and how to proceed. Then the Commission called the game.

The 1986 policy initiative

Unimpressed by the mounting opposition to an encompassing regulatory policy approach, in spite of the now blissfully obvious lack of political allies with the possible exception of the European Parliament, and in disregard of the fact that the established non-regulatory approach of DG Research and Industry was widely appreciated among the affected industries and member state governments alike, the Commission adopted a new communication to the Council (European Commission 1986). Drafted by BRIC and merely six pages long, the communication first pointed out that use of biotechnological techniques in enclosed laboratory environments and for the purpose of manufacturing processes have not been shown to pose 'any extra or new risks'. Repeatedly citing the recent Organisation for Economic Co-operation and Development (OECD) report on recombinant DNA safety considerations, which had been written in cooperation with BRIC, the communication explicitly referred to the report's conclusion that in 'those few cases in which higher risk organisms have to be used', containment measures based on existing safeguard provisions would suffice. On the issue of the field release of genetically modified organisms, it equally referred to the opinion stated in the OECD report that 'there is insufficient experience at this stage to lay down a coherent set of regulations'.

But in stunning contradiction to the position elaborated on the communication's introductory pages, the Commission statement went on to make the case for massive regulatory intervention at the EU level and found 'the rapid elaboration of a Community framework of biotechnology regulation to be of crucial importance'. It cited the industries' dependence on regulatory certainty as a rationale for this move and enumerated as further regulatory objectives the protection of consumers, workers and the environment from 'any potential hazard'. Finally, the communication stressed the necessity of harmonising regulatory regimes as part of the internal market: 'Nothing short of Community-wide regulation can offer the necessary... protection.' Accordingly, the communication announced the Commission's intention to introduce proposals for Community legislation on biotechnology by the summer of 1987. BRIC was put in charge of the drafting process. DG Industry and Environment assumed shared *chef de file* responsibility for a planned directive on the contained use of genetically modified microorganisms, and DG Environment assumed the sole responsibility for drafting a second directive on the field release and marketing of genetically modified organisms.

Decline of DG Research

In response to the new activism of the other Commission services, DG Research persistently expressed its strong opposition to what it perceived as an unjustified stigmatisation of recombinant DNA research and the danger of demonising diverse areas of development based on the scientific processes they entailed. In an attempt to reinstate itself in a central role in the process of policy formulation, DG Research organised an exceptionally well attended meeting of the Biotechnology Steering Committee in 1988. The meeting was called to discuss the growing demand for comprehensive policy init-iatives as expressed, among other places, in the resolution on the Viehoff Report by the European Parliament one year earlier. But DG Research's success in attracting unprecedented attention to the work of the Steering Committee turned against itself. The acknowledge-ment of the increasing importance of biotechnology policy and the attendance of many senior officials from different DGs only empha-sised, in the view of the participants, the need to resolve internal institutional issues first. The end of the Commission's term was only months away and the high level participants decided to postpone any decisions until after the expected reshuffling following the arrival of the incoming Commission. After this, the Biotechnology Steering Committee never met again. DG Research's plans for a high-level scientific advisory board, a last ditch effort to influence the shape of a future EU biotechnology policy, were drafted against the opposition from other Commission directorates and not pursued any further.

Biotechnology safety directives

The choice of regulatory approach

Between 1986 and 1988, when the two directives where formally communicated to the Council (European Commission 1988a, 1988b), several conflicts dominated the work of the Biotechnology Regu-lations Inter-Service Group. Most notable were disagreements about the way to break the complex issues down into manageable pieces and define the scope and applicability of the planned regulation. Should biotechnology regulation address the products or the process by which they were created? If the regulation were to address pro-ducts, the approach would parallel the vertical, sector-based, organ-isation of most relevant Commission directorates, such as DGs Industry and Agriculture, as well as the organisational logic of the affected industries and their interest representation. A process-based approach, on the other hand, would require a sector-crosscutting, horizontal pol-icy that defined the standards and regulatory rules for the application of the research technique of genetic modification and the marketing of the products based on it. Experience in the latter approach existed in

DG Environment, which had just completed the formulation of horizontal regulation in the context of a chemicals directive. Predictably, DGs Research, Industry and Agriculture were pushing for the vertical, sector-based approach. DG Environment favoured horizontal, crosscutting legislation. The political support for the traditional sector-based policy approach reflected years of experience with product legislation, in particular in DG Industry. But close ties to an industry organised by sectors, both on the Community level and nationally, which made up the Commission's political environment, also played a major role. It was through these sector-based channels that the affected industries would voice their discontent and criticise the horizontal regulatory approach, which they feared would be too interventionist and restrictive. The science community joined the biotechnology industry's opposition to horizontal legislation, albeit for largely different reasons. In the end, the choice the Commission faced was a political one. Which way it proceeded would affect everything that followed.

DG Environment was given virtually free reign over the drafting of the directives. In the process, the directorate even limited the input of the BRIC and ignored changes made during inter-service meetings over the final versions of the legislative proposals. Stressing the need to protect the environment and human health from an uncertain risks, and echoing existing unease with this new scientific technique among the public and in the Parliament, the proposals for the resulting directives emerged as some of the world's strictest. With DG Environment as the main or exclusive *chef de file* of both directives, the Commission tabled the proposals in the Council of Environmental Ministers. During their last meeting on 23 April 1990, environment ministers changed the treaty base of the directive 90/219 on the contained use of genetically modified microorganisms from Article 100a to the new Article 130s of the environment chapter in the Single European Act (SEA). It had become clear that the necessary unanimity under Article 130s was within reach. Since legislation based on Article 130s defined minimum regulatory standards, the change of the treaty base effectively turned the contained use directive into floor legislation. As a result, countries with a preference for stricter regulation, most notably Denmark and Germany, and countries with a preference for less stringent rules, such as France, did not have to be outvoted under the qualified majority rule of Article 100a. Lastly, almost all actors involved, DG Environment as well as the Council ministers, shared a common interest in shielding the sector-crosscutting regulation from competing political interests inside the Commission and the national governments. To enact the law under the new environmental chapter meant that legislative oversight would be more difficult to challenge. In the case of the deliberate release directive, the original treaty base was kept, primarily because the adoption of floor legislation would have undermined the harmonisation effect that was central to the law (see Patterson 1996: 332–333; Cantley 1995: 556).

Institutional interests played an even greater role in the dramatic changes of the text of directive 90/220 on the deliberate release into the environment and the marketing of genetically modified organisms. DG Environment and the environment ministers agreed to extend the scope of the directive far beyond what the Commission proposal had asked for. In doing so, Environmental Ministers effectively ignored the main input that DGs Industry, Agriculture and Research had made in the drafting of the legislation. Reflected in the so-called 'stop-gap' clause, the applicability of the directive had originally been limited to products not already covered by vertical, sector-based legislation. This decision to exclude products from the otherwise horizontally applicable directive was largely revised when Council ministers changed the exemption clause to pertain only to sectoral legislation that offered 'similar' risk assessment standards. By changing the wording of the clause, all existing and future product legislation was linked to the standards of environmental risk assessment in the new directive 90/220. Predictably, DG Environment as *chef de file* gave its approval. In the Environmental Council meeting on 23 April 1990, both directives were passed (Council of the European Union 1990a, 1990b). When they passed into law, the biotechnology safety directives extended DG Environment's influence over a wide range of products and effectively established a new EU policy field for which no explicit treaty base had existed only a year earlier.

Contained use Directive 90/219

As defined in Article 2 of the directive, the provisions on contained use pertain to 'any microbiological entity, cellular or non-cellular, capable of replication or of transferring genetic material' that has been altered 'in a way that does not occur naturally or by mating and/or natural recombination'. The law covers 'any operation in which micro-organisms are genetically modified or in which such genetically modified micro-organisms are cultured, stored, used, transported, destroyed or disposed of and for which physical barriers, or a combination of physical barriers together with chemical barriers and/or biological barriers, are used to limit their contact with the general population and the environment.' Important exemptions from the scope of the directive are the transport and the placing on the market of genetically modified microorganisms, which are covered by other EU legislation. For those microorganisms to which the directive applies, it specifies the risk assessment and classification schemes based on which safety measures are defined. Risk assessment entails the 'prior assessment of the contained use as regards the risk to human health and the environment' (Article 6). Records of the assessment are to be kept and made available to the competent authority upon request. Classification in two risk categories is based on the characteristics of the microorganisms and the planned research procedures. Specific containment measures follow from the classification. Prior to the use of the microorganisms, a notification addressed to the competent national authority is required. Following two procedures based on the risk categorisation, the use may either proceed after notification, unless the competent authority objects, or is blocked from proceeding until formal approval has been granted. Notification documentation includes the risk assessment files and detailed information on the personnel involved, the installations and the scale of the operation. Additional reporting provisions demand that the applicant must communicate to the competent authority any new information or changes of the planned procedures relevant to the risk assessment that might not have been available at the time of the notification. The directive also provides member states with the right to consult 'groups or the public' on any issue it deems relevant to the contained use (Article 13).

Deliberate Release Directive 90/220

The second directive, 90/220, regulates the release of genetically modified organisms into the environment for the purpose of experimental field releases or marketing. Any product containing or consisting of such organisms must first undergo experimental releases before it can be approved for placing on the market, leading to a step-by-step and case-by-case evaluation of potential risks. Other than in the case of contained use, the decision to approve the marketing of products covered by the directive involves all member states through a procedure of notification and consent. The directive applies to organisms 'in which genetic material has been altered in a way that does not occur naturally by mating and/or natural recombination' (Article 2). It exempts from the scope of the definition biological entities no longer capable of replication or transfer of their genetic material. The directive also does not apply to products derived from, but not containing genetically modified organisms. Notification and approval rules are set out separately for field release and marketing. For field releases, notification includes technical dossiers and risk assessments that are detailed in the annexes of the directive. No risk categories are specified in the directive, and accordingly all field releases have to go through the exact same notification and administrative approval procedures. As with the directive on contained use, public consultation and reporting provisions apply. Based on their assessment, the competent national authorities communicate summaries of approved applications to the European Commission, which distributes them to all member states. Member states can comment, but have no formal decision making role in the case of field releases. For the placing on the market of products falling under the provisions of the directive, applications are equally directed to the national authority of one country, but other member states may object to favourable decisions taken by one national competent authority. If member states' representatives fail to reach a decision by qualified majority in the responsible committee, the Commission can submit a proposal to the Council. Either the Council can decide by qualified majority within three months, or the decision reverts to the Commission. Article 16 of the directive further grants individual member states the right to restrict or prohibit the use and sale of approved products from their territory if they can show that they pose a risk to human health or the environment. In such cases, the Article 21 procedure applies analogously. The labelling provisions of directive 90/220 are so excessively vague that they allowed some member states to transpose them into national laws, which did not even require labels to mention whether a product was genetically modified. These provisions, or the lack thereof, unsurprisingly became the subject of heated political debate over the course of the biotechnology policy revisions. Yet the extent to which this issue would come to influence the course of political decision making was clearly not foreseen when the directive was adopted.

Failure of organised interests

Throughout the final stages of the policy process, doubts had remained both within the Commission, particularly in DG Research, and among affected interests, whether the directives would ever be adopted and become operational at all. Accordingly, the representation of interests was generally sluggish, interest organisation often slow and insufficiently cohesive, and some last-minute interventions remained without any effect whatsoever on the political choices. The input of the scientific community was hampered ever since DGs Environment and Industry had sidelined DG Research with the creation of the Inter-Service Committee. In late 1988, the Council of the European Molecular Biology Organisation undertook a series of last minute attempts to influence the Council and Commission. In a written statement adopted unanimously during an October 1988 meeting, the organisation called upon policy makers to drop the process-based approach and signalled support for European regulation based on product safety assessments. Two subsequent open letters by Europe's Nobel Laureates in Chemistry and Medicine in 1989 and early 1990 further reiterated the position that no scientific justification existed for the regulatory approach promoted by the Commission and that proposed legislation was based on non-scientific criteria (Cantley 1995: 560–561).

The affected industries were even slower to respond to the dynamics and to grasp their potential economic implications. Giving the impression of an 'interested bystander', they failed to produce 'any recognisable pattern of political action' (Greenwood and Ronit 1992: 85). Only after the Commission in 1984 convened a meeting with industry representatives to discuss the implications of biotechnology, the industry representatives concluded that they needed to improve their coordination and the exchange of information. The European Biotechnology Co-ordination Group (EBCG) was subsequently founded as an umbrella organisation of sectoral associations the following year. Since the lobby group lacked the position of a chairman or an independent budget, meetings were convened on a rotating base by members of the associations.

Cohesion proved difficult to achieve not only organisationally, but also on the substantive issues. Faced with the process-based and horizontal approach of DG Environment, the existing sector-based industrial and agricultural lobbies were often split internally on the issue. Representing a wide range of areas of biotechnological application, from agricultural and microbial food enzyme producers to animal health and traditional pharmaceuticals, the growing number of sectoral member organisations rarely found common ground and often remained gridlocked in positions representative of their divergent interests. In particular the pharmaceutical lobby, represented through the European Federation of Pharmaceutical Industry Associations (EFPIA), stalled attempts to form a more centralised lobby (Cantley 1995: 536). When the existing structure of interest organisations failed to provide the members with the necessary access to the Commission decision process, the last scheduled EBCG meeting in 1991 was simply never convened.

Given that the Council of European Chemical Industry (CEFIC) was the member organisation that chaired the lobby's sub-committee on regulatory issues and provided the industry position paper (ECRAB 1986) cited above, it must appear even more striking that organised biotechnology industries did not seem to sense the imminent possibility of the adoption of stringent horizontal safety legislation in their sector. No lessons from the earlier chemicals directive seemed to have been drawn. As the coherent and organisationally backed voice of the industries faded out, the positions of national governments, which had been decidedly industry-oriented in the early stages of the policy process, became more accommodating towards the horizontal regulatory approach pursued by DG Environment. When the Commission presented the Council of Environmental Ministers with the opportunity to pass community-wide regulations, its members pressed ahead virtually free from any pressure by nationally or otherwise organised constituencies.

Summary

By the end of the decade, the framing of biotechnology in the EU in terms of environmental safety and human protection had prevailed in several ways. Inside the European Commission, the choice of this policy focus put the Directorate-General for Environment in control and diminished the influence of other actors during the policy drafting of the first two major pieces of legislation in the field. The adoption of this policy frame was a conscious political choice. Traditional policy approaches that place greater emphasis on non-interventionist support for research and development were more established in the Community at this stage. The same is true for regulatory policy approaches that defined product-specific safety standards. But the choice for technology-based safety regulation, partially adopted under the new environmental chapter of the Single European Act, allowed the Commission to create a new policy field at the supranational level. As analysis of Commission publications directly preceding the drafting process reveals, the conflicts over inconsistent policy frames were left unresolved and turf wars persisted between the different Commission units across all levels of hierarchy. This degree of internal friction not only belies any attempt to perceive the Commission as a singular agent in any meaningful way during most of this period. The political conflicts inside the Commission also began to have an impact on the structuring of the larger political environment.

During the first years of EU biotechnology policy making, however, political interest formation outside the European Commission was curiously limited. The European Parliament had addressed the issue in an own-initiative, non-legislative report. This call for an integrated policy approach laid the groundwork for the Parliament's later involvement, but during the early years, no clear line of policy had emerged. Parliamentarians were far from forming coalitions along ideological lines or from seeking alliances with affected interests. Only around 1990, with the adoption of two major directives and with so many biotechnology interests sidelined and overridden, the pace of EU policy making accelerated. The way in which the lines of conflict formed at the European level in the following decade all but duplicated the initial policy controversies inside the Commission. The Commission, however, had already shifted its focus.

4
Policy Reframing

The new economic agenda

Political interest organisation

In mid-1989, a sense of urgency emerged among leading European biotechnology companies. A joint position paper of the five leading industrial associations from three years earlier (ECRAB 1986) had left little impact on the Commission's drive towards new safety regulation. Meanwhile, attempts at forming an organisationally more cohesive form of interest representation on the European level had remained forestalled by internal disagreement (Greenwood and Ronit 1992: 91). As the manifest failure of the biotechnology industries to wield any influence in the process slowly leading up to the adoption of the bio-technology safety directives of 1990 became increasingly apparent, the Council of European Chemical Industry (CEFIC) unilaterally proceeded to form the Senior Advisory Group on Biotechnology (SAGB) in 1989. In its original form, the lobby group combined five major industrial players in the fields of biotechnology applications, Hoechst (Germany), Monsanto, Rhône-Poulenc (France), Montedison (Italy), Unilever (the Netherlands/UK), and Sandoz (Switzerland). Based on direct member-ship, the SAGB was purposefully limited to a small group of companies. Underlining the goal of maximum internal cohesion, participation in the lobby group was by invitation only (Greenwood and Ronit 1995: 80). The group's activities were directed to the level of Commissioners and the Commission President. In fact, Commission President Delors would later famously quip that the SAGB was the most influential Euro group in existence (Greenwood and Ronit 1995: 81).

In response to the formation of the SAGB, DG Environment turned to national biotechnology associations and initiated a first meeting on issues relating to the field release of genetically modified organisms, the subject of the main safety legislation under discussion. Interpreting the move as a threat to its newly gained status as the main voice of biotechnology industries on the European level, the SAGB cancelled its participation (Greenwood and Ronit 1994: 47). The national biotechnology associations from Belgium, Denmark, France, Italy, Spain, the Netherlands, and the UK, however, established the European Secretariat of National Bioindustry Associations (ESNBA) following their Brussels meeting. This large group of associations and the predominantly small and medium-sized companies they represented added a second voice of the biotechnology industries. Given the different membership structure of ESNBA, DG Environment had created an effective venue to communicate with the national industries and thus provided an organisational structure to balance out, or bypass entirely if necessary, the views expressed through the SAGB.

Short of undermining the biotechnology lobby, the framing of the issues and the continuous competition over the policy approach between directorates inside the Commission eventually facilitated the organisation of the biotechnology industries' political interests. The affected companies had taken a long time to realise the inadequacy of the interest representation provided by the established sectoral lobby organisations on the national levels and, to a lesser extent, the European level. The fact that some countries with vested economic interests in biotechnology, such as Germany, had not even seen the emergence of genuine biotechnology associations nationally and instead continued to voice their concern through traditional sectoral organisations, probably further contributed to the slow response on the European level (Greenwood and Ronit 1995: 76–77). Where biotechnology associations existed, they often remained understaffed and insufficiently funded. Even four years after the shock of the 1990 biotechnology safety directives, only one out of eight existing national associations counted more than three permanent staff (Greenwood and Ronit 1994: 45).

With the creation of the SAGB in 1989, the representation of the interests of the biotechnology industries at the European level finally began to change. The SAGB translated informal communication networks into a sector-crosscutting lobby that could adequately monitor and respond to the policy challenges the industries faced. Its first major publication was circulated in early 1990. The positions SAGB advocated included a passionate call for product-based safety regulations in line with international regulatory standards and a call for the exemption of research and development from specific, technology-based safety restrictions (SAGB 1990a). The policy statement thus starkly opposed the approach proposed in the two draft directives. But with less than a year to go before the two 1990 directives were formally adopted, and long after DG Environment and their Council counterpart had assumed responsibilities for the legislation, the new lobby failed to wield any influence.

Economic competitiveness dimension

When it became obvious that the affected industries and the science community had lost the political battle over the legislation controlling biotechnological research processes and product authorisation, the industry lobby group changed the tone of its message. In contrast to the earlier enumeration of industries demands, the SAGB's second publication openly contested the prevailing policy frame itself. Circulated only weeks before the biotechnology safety laws were adopted, the headlining question read: 'Why can't Europe compete for commercial biotechnology investment?' (SAGB 1990b) Focusing exclusively on 'economic benefits and European competitiveness', the Senior Advisory Group had chosen to wholeheartedly ignore the predominance of biotechnology safety concerns and simply shift the focus. The new strategy proved effective. Only months after the adoption of the two biotechnology safety directives, the now manifest conflicts over the issues brought it to the renewed attention of the political and administrative leadership of the European Commission. After consultation with the Commission President's personal cabinet, the Commission Secretary General David Williamson called a meeting on the level of Directors-General. A new Biotechnology Coordination Committee (BCC), whose chairmanship Williamson personally took over, was created to deal with inter-service conflicts and general policy coordination between the Commission directorates involved in the field of biotechnology. The existing organisational structures, the Biotechnology Steering Committee and the Biotechnology Inter-service Regulation Committee, were abandoned. During the first meeting of the BCC, Williamson indicated that he intended to delegate the chairmanship of the committee to the Commission directorate that was most involved in future biotechnology policy formulation (Cantley 1995: 636). Revealingly, until Williamson's replacement as Secretary General at the end of the second term of the Delors presidency in 1995, the decision to assign the chair position to any of the DGs was never taken, and the chairmanship has remained with the Commission Secretariat-General, including after the recent reorganisation of the Commission's biotechnology committee structure in 2003.

April 1991 communication

The initial reorganisation of the Commission's organisational approach to biotechnology coincided with the publication of a communication in April 1991 on 'Promoting the competitive environment for the industrial activities based on biotechnology within the Community' (European Commission 1991). The communication was drafted by DG Industry under the leadership of Commissioner Bangemann together with the Commissioners for Research and Agriculture. It followed an even more swiftly delivered working paper by the same lead DG in the previous year. Both in tone and content, the influence of the recently found voice of the biotechnology industries was strikingly evident. More significantly, however, the April communication was part of a much broader agenda setting initiative by DG Industry, that built on an earlier communication on 'European industrial policy in the 1990s', and included a third paper on information technology. Cross-references between the texts were frequent, and the three communications were published together in a special supplement to the Bulletin of the European Communities.

The April communication declared that it was of 'paramount importance that the industries using biotechnology develop competitively', thus identifying as the Community's main objective the ability 'to create favourable conditions for the biotechnology industries, which are crucial for the development of the Community as a whole and will affect competitiveness across a broad spectrum of the Community's industries.' In this context, the creation of the BCC was portrayed as a sign of the Commission's 'recognition of the need for a more cohesive approach'. As emphasised at the outset of the section on policy recommendations, the aim to 'avoid creating undue burdens for industry' was considered paramount.

The thrust of the specific arguments advanced in the communication centred on the requirements of the Community's internal market. Existing market structures were deemed too fragmented and obstructive. Specifically, the communication found that existing and potentially envisioned regulatory provisions hamper economic development. The focus of new legislative initiatives, the argument went on, should instead be on establishing patent protection in the area of biotechnology. Citing more encompassing patent protection in other international markets, the existing Community regulations were seen as discouraging scientific innovation and economic growth, because companies remain uncertain whether and to what extent they will be able to market products based on biotechnological research. Where innovation in the Community occurred, on the other hand, a closer adaptation of existing regulations to 'international policy strategies' was said to be needed in order to provide for a more beneficial economic climate.

To achieve this goal, the communication suggested, 'a number of products will have to be regulated under existing Community sectoral legislation'. In practise, this 'efficient and simplified interaction between sectoral and horizontal legislation', was propagated to progressively reduce the scope of the horizontal directive 90/220 and limit it only to cases intentionally not covered by other laws. 'Duplications of testing and authorisation procedures will be avoided', the communication declared. The criteria for product assessment, traditionally encompassing safety, quality and efficacy, were described as sufficient. No need to account for specific characteristics of biotechnological products was noted. In particular, the communication argued throughout that the calls for a fourth criterion, designed to take into consideration socio-economic effects of such products, were unwarranted. The European Parliament had repeatedly discussed such an extension of the criteria used in product assessment and authorisation procedures. Deeming them unscientific and lacking in objectivity, the Commission communication strongly advised against their consideration.

Delors White Paper

With the April 1991 communication, DG Industry had forcefully laid claims to the issue of biotechnology and forged a strong alliance within the Commission to pursue its agenda. This initial surge was decisively boosted with the adoption of the 'White Paper on Growth, Competitiveness and Employment' (European Commission 1993a) two years later. By singling out biotechnology as one of only three policy areas explicitly addressed in the White Paper (commonly known as the Delors White Paper to reflect the importance attributed to it by the Commission President at the time), it now installed the new policy approach of DG Industry as the official Commission policy in the area of biotechnology. The White Paper heralded biotechnology as 'one of the most promising and crucial technologies for sustainable development in the coming century.' Already the sectors where biotechnology had direct impact were calculated to contribute 9 per cent of the Community's gross added value and for 8 per cent of its employment. Taking the international growth in the sector into account, the White Paper argued that the economy of the Community would see an increasing importance of the field in future years, in particular in the areas of chemicals, pharmaceuticals and agricultural processing.

In the analysis of the Delors White Paper, however, the realisation of these prospects was linked to EU policy reform. In international comparison, the Community fell behind or just compared with the US and Japan. The Community growth rate, the White Paper warned, 'will have to be substantially higher than at present to ensure that the Community will become a major producer of such products, thereby reaping the output and employment advantages while at the same time remaining a key player in the related research area'. Pointing out that there remain obstacles to achieving an 'optimum exploitation of these technologies', the biotechnology section of the White Paper roundly called for the creation of a new and 'appropriate regulatory and political environment'. In exceptionally sharp criticism of its past focus on biotechnology safety and the regulatory regime in place, the Commission went on to acknowledge that 'the current horizontal approach is unfavourably perceived by scientists and industry as introducing constraints on basic and applied research and its diffusion and hence having adverse effects on EC competitiveness.' As a result, the Commission pledged 'to be open to review its regulatory framework'. Only by harmonising EU biotechnology safety regulations with 'international practice' could the Community ensure that it will become more than just a market for biotechnology-derived products. The policy recommendations put forth in the Delors White Paper consequentially stressed 'flexibility and simplification' of risk assessment and market approval procedures. Furthermore, the White Paper called on the Commission directorates to 'reinforce and pool the scientific support' for such regulations in such a manner as to give the scientific community a strategic role to play in the development of future Community policy in the field of biotechnology.

Commission follow-up communication

Under the direct political leadership of the Secretary General, the Biotechnology Coordination Committee was subsequently charged with drafting a follow-up paper to be presented at the next European Summit in June 1994. This communication, entitled 'Biotechnology and the White Paper on Growth, Competitiveness and Employment – Preparing the next Stage' (European Commission 1994), already echoed the new call for a stronger emphasis on the economic potential of biotechnology in its introductory passages. From this reassertion of the new policy focus, the paper directly moved on to assess the existing regulatory framework from this perspective. 'Put simply', one independent news agency wrapped up its lengthy report on the technical changes envisioned in the communication, 'the Commission wants to ease the current legislation' (European Information Service, 14 June 1994).

In its communication, the 'Commission acknowledged that the biotechnological regulatory framework is a factor impacting on industrial competitiveness'. The existing regulatory regime, set up primarily through the two 1990 directives, 'has been built upon the knowledge available at that time, when there was still considerable uncertainty as to the safety and risks involved in the application of modern biotechnology'. As a result, existing legislation was characteristically 'aiming at a broadly preventive approach' as regards the technology's application. Taking into account the objective of economic growth, however, the Commission 'confirms the need for balanced and proportionate regulatory requirements' as part of the more 'adaptable regulatory system' that it envisioned. Towards this end, the communication outlined a two-track approach of revisions and reforms. This approach would entail for the Commission 'to exploit fully, where they exist, the inherent possibilities to adapt [the biotechnology regulatory framework] to technical progress (via regulatory procedure). At the same time, it will bring forward amendments in order to incorporate changes which cannot be achieved by technical adaptation while leaving the basic structure of the framework intact.' Changes to directive 90/220 on the deliberate release of genetically modified organisms were announced, to be presented after a review scheduled for 1995. However, the more specific section on the law clearly stated the goals: 'extending the flexibility of Directive 90/220/EEC, so that its scope and the procedures to be followed are always appropriate to the risk involved, and are easily adaptable', furthermore 'facilitating the link between this Directive and product legislation'. Two legislative proposals under discussion at the time, and referred to in the text in several contexts, were singled out as 'matters of urgency': the planned regulation on Novel Food and the directive on patents for biotechnological inventions. As in the earlier White Paper, the patent directive in particular was portrayed as a major cornerstone of the new approach to biotechnology and economic competitiveness in the Community.

In light of these developments, the industry lobby SAGB was brimming with confidence. Responding to the publication of the Commission follow-up communication, the lobby organisation published press releases that demanded that the Commission move 'further and faster' (European Information Service, 10 June 1994). The statement to the press in particular emphasised agricultural biotechnology as an area of great promise, and it called upon the Commission to extend its strategic review beyond the existing regulatory scheme and instead include all pending or proposed legislation in a general overhaul of EU biotechnology policy.

Summary

After the adoption of the two biotechnology safety directives in 1990, the Commission starkly shifted its policy focus. All subsequent Commission publications on the subject advocated substantial revisions to the regulatory regime just put in place. While the focus of the first biotechnology directives had been almost exclusively on human safety and the protection of the environment from existing and unforeseen risks associated with a relatively new scientific technology, a new emphasis on economic competitiveness and growth was now the overriding concern. With the shift in focus, other political actors and policy venues gained in importance. Primarily, the shift empowered the Directorate-General for Industry and it effectively sidelined DG Environment for the better part of the early and mid-1990s. The

European Commission's diverse interests in the policy area of bio-technology and the conflicting ways in which competing fractions inside the Commission addressed and pursued the issues were also beginning to affect the political environment at the European level. Political interest representation emerged and changed in response to adopted policy and direct requests by the European Commission to form more effective lobby groups. The divisions inside the Commission had even translated into the structuring of interest group representation, and two major competing organisations with mutually exclusive membership had emerged. The initial conflict over the framing of the issues had significantly gained in scope and organisational heft.

A new legislative agenda was now taking shape, both in terms of increasingly specific proposals for reform and in terms of the larger, more obvious, shift of the policy context in which the Commission addressed the issues and re-evaluated the existing laws. Yet despite the dramatic turn-around propagated under the new policy frame, the Commission publications from this period were short on comments or analysis concerning the political ramification of the Commission's new policy focus. Where political aspects were taken up, the analysis was at times strikingly implausible. References to disadvantageous public perceptions are a case in point. The 1994 follow-up communication, for example, took the especially pointed view that strict regulatory regimes may increase public hostility towards biotechnology. However, the general impression was conveyed that public education about the low level of risk implicit in biotechnological applications would eventually help to overcome reservations. In two short passages, both in the introduction and in the conclusions of the follow-up communication, reference was also made to the European Parliament. The Commission, it stated, 'recognises the important interest of the European Parliament in developments in biotechnology and is ready to establish the necessary dialogue on biotechnological issues', both with Parliament and through continuing its tradition of holding Round Tables with industry representatives and other interest organisations. Following the communication, the Commission was called upon by the Council of Industry Ministers in a meeting in September 1994 to prepare proposals for a revision of the Contained Use Directive 90/219 and to table them at the European Summit held the end of that year.

The Industry Council ministers also called upon their counterparts in the Environment Council to address the issue 'without delay'. Yet the environment ministers did not even place the topic on their agenda (see Cantley 1995: 652). Parliament, to the contrary, was more than ready to deal with the issue.

Policy contestation

With the adoption of the follow-up communication, the Commission had chosen an awkward double strategy. Since the enactment of the two biotechnology safety directives, three consecutive Commission policy statements had established economic competitiveness as the new focus, dominating both the titles and the substance of its policy recommendations. This shift of focus seemed to put DG Industry in charge of policy formulation and regulatory review. But the Commission strategy also foresaw that it would pursue the necessary revisions as far as possible by way of internal amendments and only resort to proper legislative action if the planned changes could not be pushed through under the articles of the two 1990 directives, which allowed for the 'adaptation to technical progress'. The procedural strategy clearly conflicted with the objective to rewrite EU biotechnology legislation from a completely new angle.

In the case of the contained use directive, internal amendments could be made through the so-called 'Article 21 Committee', named after the corresponding article in the directive, which was composed of member state representatives and chaired by the Commission. Not surprisingly, the supporters of the original safety frame did not view the committee or the idea of internal revision favourably. In September 1993, during a roundtable on the biotechnology regulatory framework organised by Commission Secretary General Williamson in his function as the chairman of the Biotechnology Coordination Committee, the environmental interest group Friends of the Earth submitted written comments in which it attacked the Commission strategy as straightforward safety deregulation. 'Although the expression "deregulation" is always avoided when discussing the "necessary adaptation" of the EC regulation concerning genetic engineering to the state of the art of scientific risk assessment', the statement explained, 'the so-called simplification is in fact a deregulation because it had been initiated to decrease central recommendations on how to proceed with risk assessment, although neither the methodology nor the data for a reliable prognosis, especially on the long term ecological effects of GMOs, are available so far' (quoted in European Commission 1993b).

As an internal document (European Commission 1993b) circulated by the Commission Secretary General after the Round Table discussion reveals, the exchange of views that followed the presentations of high-ranking Commission officials reflected greatly divergent views on virtually all issues under policy review at the time. The Community's economic competitiveness was raised first and figured most prominently during the discussion. It became clear, however, that the diametrically opposed analyses of the participants trumped all hopes of establishing common ground any time soon. While some participants welcomed the new approach as long overdue and supported the more specific policy reforms envisioned in past Commission publications, resistance was voiced as well. The emphasis on the regulatory regime as a determinant of Europe's competitiveness in biotechnology was deemed simplistic. Proposals for simplified risk assessment and product approval procedures were instead characterised as potential 'dilution of the assessment of the risk'. Far from supporting any type of deregulation, existing safety provisions were considered insufficient on the grounds that they inadequately reflect concerns for possible long-term effects on the eco-systems and because they were still based on very little actual experience concerning the future consequences of biotechnological applications, particularly in agricultural usage.

Given the evident state of disagreement during the Round Table discussion, the participants left the Secretary General with few options other than to conclude the meeting by reiterating the present Commission position on biotechnology policy. Among his points, such recurring themes as the planned shift from horizontal to sectoral product legislation, that incorporated its own risk assessment procedures, stands out as conflicting with the general line of criticism voiced before. Beyond an exchange of views over biotechnology policy reform, little if any common ground was found. Environmental interest groups such as Friends of the Earth and Greenpeace were regular participants during the Commission Round Tables on biotechnology. But their invitation and participation amounted to little more than participatory window-dressing, as senior Commission officials freely acknowledge (Interview, European Commission, March 2005). The new policy agenda was all but set, the respective positions were known in advance. For environmental groups these developments meant that their allies inside the Commission were getting fewer. As the events unfolded, however, it became clear that the political leverage of EU environmental interests was far from diminished.

The rise of Parliament

Parliamentary reports on biotechnology

In response to the new economic focus in biotechnology policy expressed in the Commission communication of April 1991, Parliament had decided to write a new own-initiative report on biotechnology in keeping with the tradition of the Viehoff Report of 1986. Hiltrud Breyer (Greens, D) was assigned the rapporteur function. In tone and substance, the draft report (European Parliament 1994) of her committee was a frontal attack on the Commission and its new policy frame, and a loud and clear signal to all affected parties opposing the shift in policy focus to close ranks with the Parliament on matters of biotechnology safety. The report called on the Commission 'to state openly that genetic engineering raises fundamental environmental and health concerns and that therefore strong measures to protect human health and the environment are a fundamental element in the development of any activities in genetic engineering and when considering competitiveness'. It went on to defend the horizontal safety approach adopted by the Commission in the two 1990 directives and called on the Commission to 'discontinue its practice of proposing sectoral legislation which transfers environmental risk assessment of the marketing of genetically modified organisms (GMOs) out of directive 90/220'. In a rare attack on the Commission Secretary General Williamson and his role in the implementation of Commission President Delors's agenda, the report fiercely called upon the Commission to 'ensure that in administrating the Biotechnology Coordination Committee, the Secretariat-General does not overstep its proper function of coordinating the Commission services and take on a policy role in this area.' The report further presented 'factual evidence that legislation on safety and product assessment in the EU does not place industry here at a competitive disadvantage vis-à-vis the United States and Japan', thus fundamentally questioning the validity of the new Commission approach. Instead, policy initiatives advanced by DG Industry were accused of 'singling out specific industries for corporatist protection from competition'. Directed at the member states, the draft report insisted that Council 'reconsider its hasty approval of the Commission communication on Biotechnology in the light of the above concerns' and, more than everything else, 'maintain the integrity of directives 90/219 and 90/220 and not accept any proposals to deregulate'. Finally, the report called on Parliament to commit itself to 'reject any proposal from the Commission to amend directives 90/219 and 90/220 or to weaken them through internal revision or transfer of authority'. Parliament was determined to block both paths of the Commission's newly developed 'two-track approach' of adaptation to technical progress, which in the committee's view was nothing but barely concealed policy revision outside the channels of proper legislative decision making and actual legislative review.

The draft report coincided with the Commission's adoption of the follow-up communication on biotechnology and the Delors White Paper. It was eventually adopted only in 1996 as a direct reply to the formulation of the new Commission agenda (European Parliament 1996a). While the thrust of the report changed little, many of the direct attacks on the Commission and Council in the draft report were dropped or rewritten during this period. The process of deliberation over the draft report in Parliament, however, proved informative for a very different reason. While the Committee on the Environment drafted the report, as had been the case with the Viehoff report ten years earlier, standard parliamentary procedures allowed other committees to formulate opinions, which were then attached to the draft and circulated along with it. As the opinions of other committees reveal, strikingly little internal competition over the issue of biotechnology and the regulatory reform existed inside the European Parliament at this time. The Committee on Economic and Monetary Affairs, and Industrial Policy would have been the obvious parliamentary arena for proponents of a new framing of biotechnology to pursue their policy agenda. The draft of the report by the Environment Committee had put an unyielding focus on environmental safety. Now, the review of the biotechnology policy in light of the Delors White Paper on Growth, Competitiveness and Employment was clearly the best platform that advocates of a more industry-oriented policy inside Parliament could ever hope for. But the respective parliamentary committees either dropped the ball or never saw any reason to contest the framing of the issue and the corresponding dominance of the Environment Committee. When the Economic and Industrial Affairs Committee unanimously adopted its opinion on the report 21 June 1995 (European Parliament 1996a), the committee clearly failed to push for the new focus on economic competitiveness in biotechnology policy that the Commission had so aggressively pursued. Instead of taking the side of economic interests, the committee opinion notes that on the issues of EU biotechnology policy 'the various needs and views in play make it difficult to arrive at a broadly acceptable position'. Far from arguing for the importance of any one of these conflicting approaches, the committee members went out of their way to acknowledge various underlying policy interests such as consumer protection, safety, and the environment. In passing reference to the core arguments of the Delors White Paper the opinion wearily acknowledged that 'the field of biotechnology may well have a positive effect on employment'. Yet the role of biotechnology in the context of Europe's international competitiveness as an economic actor is addressed only briefly. The committee's lack of impetus in pursuing any distinguishable policy agenda of its own is as striking as its apparent lack of ties with the biotechnology industries at the European level. In the last point of the conclusions, the committee opinion advocated the establishment, among others, of socio-economic criteria to assess biotechnology in the EU – an idea that the affected industries had been vehemently fighting against for years in their feverish attempt not to hand the Commission yet another yardstick against which to judge the risks and promises of biotechnological innovation. Eventually, the line was adopted into the final version of the report. The members of the Committee of Economic and Industrial Affairs must have been either blissfully unaware of the ongoing policy debate, or they acted in complete disregard for European biotechnology industries interests. The first interpretation does not seem entirely not out of the question, as the text of the committee's opinion also confused the titles of the biotechnology laws under consideration.

Beyond the adoption of a call for socio-economic criteria as part of the evaluation of biotechnological applications, the final report by Parliament (European Parliament 1996a) confirmed its outright opposition to the new economic competitiveness frame under which the Commission had begun to discuss and promote biotechnology reform in the European Union. In the report, the Parliament adamantly insisted that legislative reform of the existing regulatory framework should aim to increase regulatory certainty and must not reduce harmonisation or compromise in any way the primary objective, the 'principle of maintaining safety for people and the biosphere'. The only way to do so, the report concluded, was by building on the horizontal safety legislation, not on specific product laws. In particular, environmental risk assessment that lies at the core of the early directives from 1990 must remain the yardstick for possible specific sectoral legislation and 'must be retained as catch-all provisions'. In complete opposition to the Commission's agenda, the Parliament further included calls for 'comprehensive labelling of genetically modified products, including seeds', and urged the Commission 'to extend product liability legislation to goods and services issuing from genetic engineering'. Parliament had most certainly not shown any signs that it was willing to yield ground in the battle over the future of Europe's biotechnology policy. It became clear that the reform of the biotechnology regulatory framework envisioned by the Commission under the new frame of economic competitiveness was impossible to implement through regulatory committee decisions alone. If policy reform was to be pursued by way of proper legislative action, Parliament would be involved, and increasingly so with every treaty reform. The extent, however, to which the conflict between the conflicting frames of biotechnology policy would capture EU institutional politics was difficult to foresee.

Biotechnology patents directive

At the time of the deliberation of the report in Parliament, no legislative proposal for amendments to either of the two biotechnology safety directives 90/219 and 90/220 had been brought forward by the Commission. Another major piece of biotechnology legislation, however, was in fact under discussion in the Parliament: the proposal for a directive on the legal protection of biotechnological inventions (European Commission 1988c). Since its drafting in 1988, the biotechnology patents directive had become a central component of the new approach to biotechnology and industrial competitiveness, much in line with the demands of the Senior Advisory Group and the advocacy of the Commission communications of 1991. Drafted by DG Internal Market, some member states had initially expressed reservations about the original text of the legislative proposal for a combination of legal and substantive reasons, and plans for a revised draft emerged shortly after the proposal had been communicated to the Parliament and Council. Despite this, internal coordination in the Commission was significantly less contentious than in the case of the biotechnology safety directives. DG Agriculture had initially protested against some provisions of the first draft, but as agricultural interest groups failed to mobilise, DG Agriculture shifted its attention to the drafting of a separate directive on plant variety rights that settled most disagreements with other Commission services and pre-empted overt conflict. With little conflicting pressures from inside the Commission, the biotechnology industries' lobby groups, and the Senior Advisory Group chiefly among them, pressed successfully for their interests. In many ways, the first biotechnology patents proposal was a textbook case of a market harmonisation directive. Differences in national industrial property laws had 'direct and negative impact on Community trade and there is no other field of technology where national patent laws vary on so many points as they do in biotechnology', the original Commission proposal argued (European Commission 1988b).

In Parliament however, involved under the cooperation procedure, feverish attempts by the responsible Legal Affairs Committee to reconcile different views and reservations were rebutted by delay strategies in the General Assembly. The report on the legislation was referred back to the committee twice before it was finally adopted in October 1992. Conflicts emerged most strongly over the issue of the patentability of human body parts. Underneath the level of substantive policy disagreements, however, a second and increasingly open institutional conflict between the Parliament and Commission gave the exploitation of delay tactics a second dimension. A passionate intervention by the chairman of the Environment Committee in favour of the second referral of the parliamentary report back to the Legal Affairs Committee hinted at the frame of mind under which leading Green members of parliament perceived the issue. In December 1993, the legal basis of the biotechnology patents directive was changed in accordance with the Maastricht Treaty and the legislative procedure changed to co-decision, giving the European Parliament and the Council coequal legislative powers. The political process stalled again and the directive ended in the Conciliation Committee. After two meetings, a compromise text was hammered out. But on 1 March 1995, when the directive on the legal protection of biotechnological inventions was up for vote in the assembly, the European Parliament, in an unprecedented move, vetoed the text agreed upon at conciliation and terminated the legislative process. Preceding the vote, intense lobbying efforts by environmental groups had been mounted, spearheaded by Greenpeace and their European Campaign on Biotechnology Patents. Environmental interests had been unmatched by the less coordinated and infinitely less forceful lobbying attempts by the biotechnology industries, which largely refrained from making their case before the members of the European Parliament at all. After the veto, the Senior Advisory Group laconically reiterated their standard line and argued that the vote 'sends a negative political signal to investors that will further reduce the attractiveness of the Union' (Agence Europe, 2 March 1995). Along similar lines, the European Federation of Pharmaceutical Industry Associations (EFPIA) warned in a statement 'Europe has taken another step towards being nothing more than a consumer of innovative health-care products and not a supplier on the world markets' (European Information Service, 6 March 1995).

On 13 December 1995, just months after Parliament had vetoed the first law, Internal Market Commissioner Monti announced that the Commission had adopted a new proposal and submitted it to the Parliament. The draft text addressed the most central concerns expressed by the legislators and adopted language that effectively prohibited patents on human life. Bodies and parts of human bodies were excluded from the clauses granting patent protection by introducing a legal distinction between 'invention' and 'discovery', and by extending patent protection only to the former category. The patentability of animals was restricted to cases in which the benefit to humans outweighs the suffering of the animal. The Commissioner argued that 'the right balance between the equally essential requirements of promoting research and providing ethical protection' (European Voice, 14 December 1995) had been struck, and he showed great confidence that the directive would pass the second time around. Industry lobbies engaged members of parliament in regular discussion events, outlining the industries' willingness to exert self-constraint. The way in which the European Parliament addressed the new proposal for a directive was remarkable in two ways. The first concerns the great vigour with which the Parliament took on the issue. The Legal Affairs Committee, which held primary responsibility, discussed the reintroduced proposal during over a dozen meetings in the first months following the reintroduction of the legislative proposal by the Commission alone. An exceptionally high number of committees took part in the vetting of the legislative proposal and its deliberation with constituencies. In addition to the Legal Affairs Committee, five other committees were charged with submitting opinions (see Earnshaw and Wood 1999: 297). Yet the members of the Parliament not only consumed uncommon energy in dealing with the topic, they also acted with exceptional discipline on several levels. Few aspects of the massive effort Parliament put into the deliberation of the draft law seemed internally controversial. The Green group secured for itself three out of the six rapporteur positions in the committees involved, giving it substantial influence over the internal decision making process (Earnshaw and Wood 1999: 298). Their influence, furthermore, seemed largely uncontested, and the Green group managed to push through identical proposals for legislative amendments in several committees simultaneously. In first reading, the plenary adopted all amendments proposed in the report by the Legal Affairs Committee on 16 July 1997 with a clear 370 to 113 majority and only 19 abstentions. The European Commission adopted the amended proposal virtually without changes and actively liaised between the Parliament and Council in the following months to broker a compromise. When the Council adopted a common position in February 1998, the Commission knew that the Council would not only endorse the modified Commission proposal, it also accepted all further changes proposed during Council proceedings and resubmitted the text to Parliament. In second reading, the Green group's proposal for the rejection of the common position failed. The almost desperate attempt to derail the legislative process over biotechnology patents a second time suffered a crushing defeat with only 78 votes in favour, 432 against and 24 abstentions. The rapporteur from the Legal Affairs Committee, who had already called upon the assembly to adopt a compromise draft in 1995, lobbied for the law a second time. He pointed out the concessions

Parliament had won on the issue of the patentability of human and animal life and reiterated the importance of patent rights for the growth of the biotechnology sector, especially in the case of the pharmaceutical industry. Both opposing camps feverishly continued to lobby for their cause (European Voice, 17 July 1997 and 20 November 1997).

In the days leading up to the vote, parliamentarians were targeted by a campaign unparalleled in the history of the institution to that date. 'For example', one advisor to the Green group in Parliament recalls, 'it was impossible for my office to use our fax. All the time, night and day, we received faxes from the industry lobby' (Interview, European Parliament, March 2005). Environmental pressure groups employed much the same strategies. The Brussels-based Campaign on Biotechnology Patents increasingly took over the coordination of activities, building on a network of highly professional activist groups. One such group, Global 2000, started its campaign against the directive 'by sending twice-weekly faxes to about 250 MEPs on the key six committees' (European Voice, 3 July 1997) before it proceeded to establish personal contacts with key members of Parliament. Yet environmentalists remained less visible compared to the impressive public campaign that groups like Greenpeace had staged three years earlier. Unable to claim the moral high grounds, their tactics seemed toned down in realistic anticipation of a swing of the political mood. As one rapporteur noted, most direct contacts during this time were with the industry lobby, animal welfare groups and organisations representing the rights of patients. In the end, this last group was critical for the direction of policy deliberation. Outside Parliament, patients' rights groups such as the European Alliance of Genetic Support Groups in yellow T-shirts dominated the scene. Eventually, the presence of large numbers of actual patients suffering from genetic diseases drew by far most of the attention. As one commentator observed, they changed the 'parameters of the debate' and 'added an extra dimension' (European Voice, 3 July 1997). Their message for the members of the European Parliament was that they relied on the biotechnology patents directive to provide the industries with the legal basis and the economic incentives to invest in the research that their lives depended on. 'The coalition of patients' rights NGOs with industry, the whole concept of progress', one observer from the Green group reflected glumly, 'all of that was finally a very magic mixture' (Interview, European Parliament, March 2005).

On 6 July 1998, the directive on the legal protection of biotechnological inventions was passed into law (European Parliament and Council of the European Union 1998). The biotechnology industries had woken up to the implications of the Maastricht Treaty and had engaged, as the chairman of the Senior Advisory Group had pledged immediately after the 1995 veto, in a dialogue with a 'very wide audience' (Agence Europe, 24 August 1995) that now even included both chambers of the legislature. But the environmental lobby was far but defeated and first signs quickly emerged that Greenpeace had all but left the fray. The organisational basis and expertise gained through the European Campaign on Biotechnology Patents was re-deployed and facilitated an even more vigorous campaign against the release of genetically modified organisms. This campaign took shape when it became public in late 1996 that genetically modified soybeans had entered the European market without proper labelling (see van Schendelen 2003: 227).

Controlling scope and direction

Revising the contained use directive

Before the battle over the different frames in biotechnology policy cumulated in the repeal of the deliberate release directive 90/220, very different political dynamics characterised the amending of the second safety directive from 1990 on contained use and laboratory research. Eventually, the proposal for amendments to the contained use directive (European Commission 1995) as envisioned in the 1993 and 1994 communications and endorsed by the European Summit and Council meetings in 1994 led to a highly publicised fall-out. Due to the change of the treaty base shortly before its adoption, amendments to directive 90/219 were based on Article 130s (SEA), thus limiting the influence of Parliament under the cooperation procedure. With the weight of the White Paper on Growth behind its plans for policy revision, and after having detailed the envisioned modifications in the follow-up communication of 1994, the ball was in the Commission's court and steps towards deregulation and more industry-friendly safety provisions seemed well within reach. Industry had been pushing hard for the adoption and implementation of changes through the Senior Advisory Group. Additional pressure was exerted during high-level interventions from German industry and the affected research community. During the first month of the German presidency in 1994, a group including the President of the Max Planck Society and representatives of the German chemicals industry were received by Commission President Delors to discuss a list of changes to the two existing biotechnology safety directives. The pharmaceutical and chemicals industries' particular concern with the contained use directive was that it covered processes in which products are derived from or processed with the help of genetically modified microorganisms, but the products do not consist of them in the form placed on the market. Since much of the chemicals industry applied such procedures, the research and processing facilities of these industries were severely and adversely effected by directive 90/219. As a result, they considered that standard scientific procedures were over-regulated, due solely to their use of genetically modified microorganisms and indiscriminate of the level of risk involved in the processes. The general demand for reform was further underlined by a resolution on the legal framework for genetic engineering, adopted by the European Science Foundation in the same year, which called upon the Commission to revisit the original formulation of risk categories, clarify risk assessment based on scientific knowledge and simplify the approval procedures for limited field trials (Cantley 1995: 655–656).

When the revision of the directive on contained use came up for discussion however, DG Environment was *chef de file*, unlike during the process of the drafting of the original directive, when DG Industry had shared the formal responsibility and co-authored the text of the law. The Environment Commissioner shared none of the enthusiasm for safety deregulation, and the process of policy coordination within the Commission predictably broke down almost completely. Eventually, German Chancellor Kohl addressed the issue in a public letter to the new Commission President Jacques Santer. In the letter, the Chancellor reiterated the 'common goal to encourage growth, competitiveness and employment in Europe' and criticises that in this context, the delays to the revision of the contained use directive by DG Environment were unacceptable. The exchange of German government officials with Environment Commissioner Bjerregaard was said to offer 'little comfort, creating the impression that the topic is not a Commission priority' (BioTechnology, February 1995). Inside the Commission, Commissioner Bjerregaard cut off contacts with the Biotechnology Coordination Committee, including with the committee members from her own Directorate-General. Instead, she charged her cabinet with the task of producing an independent second draft. Among other members of the BCC, work on the earlier draft continued and the cooperation between DGs Research and Industry markedly intensified. Problems surfaced when it became evident that DG Industry lacked enough experienced staff in this highly technical policy field to match the expertise of Bjerregaard's cabinet. Looking for external assistance, DG Research reached out to the German Ministry for Health. The ministry promptly seconded an expert on the issue to DG Industry who immediately took up the task of drafting an amendment in close cooperation with the respective Commission expert in DG Research (Interview, European Commission, March 2005). On the formal level, a BCC *ad-hoc* working group composed of national representatives and Commission personnel preceded to hammer out details in closed-door sessions (Patterson 1996: 336, 342). When Bjerregaard tabled her proposal for the revision of directive 90/219 for adoption by the College of Commissioners on 6 December 1995, the version was voted down and Industry Commissioner Bangemann presented the BCC counter proposal. Against only two votes, including the Environment Commissioner's own, the college of the Commissioners adopted the BCC proposal and communicated it to Council and Parliament.

Industry interest groups welcomed the news of the uncharacteristically favourable adopted draft. The Commission proceeded to push for the adoption of the amendments during a high-profile interinstitutional conference on biotechnology (Agence Europe, 10 January 1996). Revealingly, with four Commissioners present, not Bjerregaard but Commissioner Bangemann delivered the opening remarks during the conference and spoke on behalf of the Commission President, who was absent to attend the funeral of François Mitterrand. The substance of the speech was a rehash of the key messages of earlier Commission communications, following in the tradition of the Delors White Paper (European Information Service, 23 January 1996). The Commissioner formally in charge of the portfolio, Environment Commissioner Bjerregaard, stepped in later to elaborate on the regulatory details. The draft directive amending contained use directive 90/219 maintained its original horizontal character, but linked the notification procedures and containment requirements to completely revised risk categories, streamlined the administrative procedures and lowered the legislative threshold for future revisions. Under the new rules, four, instead of the originally two risk categories applied, in line with both the rules established by the World Health Organisation and analogous to the risk categories already put in place at the EU level by an earlier 1990 directive on workplace safety. For the first two risk categories, notification and containment provisions were considerably scaled down. Only in the cases of the two higher risk categories, prior written consent from the competent national authority was required (see also Sheridan 2001: 21–28). Despite the apparent aim to deregulate the existing safety provisions, the Environment Committee in Parliament proposed only modest changes to the language. Following the lead of rapporteur Trakatellis (EPP, Greece), a former biochemistry professor, Parliament voted not to adopt a large number of proposed amendments which were deemed too stringent, bureaucratic and potentially burdensome for industry. Parliament insisted however, that the Commission change the treaty base to Article 100a, thus giving Parliament more influence under the co-decision procedure. The change of treaty base was predictably denied. Council and Commission both adopted the majority of the more technical amendments either completely or partially. The amendment to incorporate a liability clause, however, was not adopted in the final version. The liability clause foresaw that legally responsible users of genetically modified microorganisms had full civil liability for any damage to human health and the environment. Accordingly, research laboratories and other applicants would have to take out sufficient liability insurance to cover any possible losses or damages. The common position subsequently adopted by the Council changed so little in terms of the substance of the revised Commission proposal that the Commission accepted it, even welcoming some technical and legal clarifications which it considered strengthened the proposal. Parliament reintroduced some of its original 57 amendments, 29 of which had already been incorporated in the final version. But Council passed the law after the Commission forwarded the re-examined proposal with no further changes in October of 1998. The central parliamentary amendments not included in the final text concerned legal and procedural issues such as the treaty base and applicable committee procedures in the law's implementation. While these issues carried considerable weight, the compromise between the institutions of the EU was almost complete on another level. Concerning questions of substantial policy changes, the revision of the contained use directive stands out as exceptionally uncontroversial. Even two years later, when the Commission set out to promote sweeping changes across the entire EU biotechnology policy regime in a White Paper published in 2000, not a single of the over 80 policy proposals included in the White Paper related to the issue of contained use or the amended directive 90/219.

Interest group reorganisation

During the period of the deliberation over the revision of the contained use directive 90/219, industry representation at the EU level underwent a major restructuring. As discussed above, early attempts to coordinate national biotechnology federations through the European Biotechnology Co-ordination Group (EBCG) had failed in 1991 and led to the formation of the Senior Advisory Group on Biotechnology (SAGB) while a number of national associations had formed the European Secretariat of National Bioindustry Associations (ESNBA). The SAGB in particular, based on limited direct membership and profiting from a high level of cohesion, was for some time considered the model of a successful European lobby group. But the case of the contained use directive made evident that the link to national political arenas remained important, and that the SAGB lacked this connection in organisational terms. Moreover, EU directives needed to be transposed into national law, which caused continuous demand for local information that the Brussels office of the SAGB had more difficulty obtaining than the national associations did. This problem was especially pressing in the case of floor legislation, such as the contained use directive, which allowed member states to exceed the safety standards of the EU provisions (Interview, European Commission, March 2005). Furthermore, the success of DG Environment in installing the ESNBA as a competing organisational structure for biotechnology interests at the EU level had undermined the role the SAGB had sought for itself. In 1996, following lengthy talks, the two lobby groups merged to form EuropaBio (European Information Service, 27 November 1996). The new organisation encompassed over 500 biotechnology companies and initially eight national associations, rising to ten by the following year and 25 by 2004. EuropaBio simultaneously functioned as the secretariat of the Forum for European Bioindustry Coordination (FEBC), which consisted of sectoral industry groups covering issues as diverse as pharmaceuticals, animal health products, food, feed additives, seeds, plant protection products, enzymes, chemicals and diagnostic products.

Pyrrhic victory: The novel food regulation

The comparatively swift and uncontroversial revision of the directive on contained use discussed in the previous paragraphs proved to be the exception. Parallel developments in EU biotechnology policy concerning different legislative initiatives during the 1990s suffered a very different fate. One law in particular exemplifies how the shift from a focus on human safety and environmental protection towards a more industry-oriented biotechnology policy raised conflicts across the EU institutions that endured over a very long legislative process. This legislation concerned the regulation of novel food, including genetically modified food products. The Commission pursued the regulation on novel food as a major step away from the technology-based safety directives, but the compromise reached after a hard-fought legislative battle eventually failed to withstand the renewed reframing of biotechnology policy in 2003 and the law came under attack almost as soon as it took effect. While the two laws discussed in the previous sections of this chapter, the biotechnology patents and the revision of the contained use directive, provide critical insights into the political dynamics of the policy field, their effects on the further legislative evolution were limited. The novel food regulation, on the other hand, played a very prominent role to the extent that it triggered the political dynamics leading to the eventual defeat of the economic competitiveness frame.

In the absence of a single coordinated policy approach during the early and mid-1980s, DG Industry, in close contact with the affected national industries, had continued to pursue harmonisation of product legislation as envisioned in the 1983 Commission strategy paper. The regulatory strategy was to exempt more and more products from the strict provisions of directive 90/220 by introducing sectoral legislation that limited the scope of the catch-all horizontal directive. Following the adoption of a first Commission proposal for a regulation on the marketing and labelling of novel food and food ingredients in June 1992 (European Commission 1992), the European Parliament delivered its first opinion in October 1993, just weeks before the procedure was changed to co-decision under the Maastricht Treaty. In its report and the following discussion in Parliament, the most controversial point concerned the issue of whether or not genetic modification of food and food ingredients demanded mandatory labelling that exceeded the general rules for labelling outlined in the proposal. The Commission draft foresaw labelling provisions based on the criterion of 'substantial equivalence', which meant that labels were only mandatory if the nature of the food product differed from an organic equivalent in ways affecting aspects such as consumption, nutritional values or allergenic substances. Linked to the notion of 'substantial equivalence', was the exemption from the full author-isation procedure. Many food products falling under this category were derived from genetically modified organisms but no longer contained them in the form in which they were sold to consumers. Based on this argument, the Commission draft proposed to introduce a simplified notification procedure that did not include the risk assessment and authorisation procedures that would have applied under horizontal law (European Commission 1992). In response to the criticism in Par-liament, the Commission adopted an amended proposal, but the pro-cess came to a standstill for almost two years. While the approach to labelling remained hotly contested between the Parliament and the Commission, the situation soon took a decisive turn towards open confrontation along entirely new lines. In November 1995, Environ-ment Commissioner Bjerregaard surprised her colleagues by unilater-ally changing a draft decision on the authorisation of a new strain of genetically modified rapeseed, after member states representatives had already approved it, to include compulsory labelling provisions. Neither the directive on the deliberate release of genetically modified organ-isms, on which the approval procedure was based, nor the Commis-sion communications following the adoption of the safety directive, included any plans for labelling of the scope now proposed by Com-missioner Bjerregaard. Angered Commission officials observed that 'she is basically on her own' (European Voice, 16 November 1995). Yet that was about to change.

While the authorisation of seeds followed legal procedures not directly linked to the planned regulation on Novel Food, the step had sent a strong signal to Parliament. Only one year after the

Commission had issued a coherent and ambitious follow-up communication on the implementation of the Delors White Paper
on Growth in the area of biotechnology, the divisions inside the
Commission had resurfaced yet again. Around the same time,
the common position the Council adopted in October 1995 on the
novel food regulation was met by strong opposition in the environment committee in Parliament. Rapporteur Roth-Behrendt (PSE, D)
expressed her dismay over what she perceived to be a failure of the
Council of Industry Ministers to take consumer interests into account
and advanced a line of attack that was ultimately in complete disagreement with the notion of product legislation as such. She argued
that product safety depended not on the type of food, but also on
the specific process of genetic engineering. In the opinion of the
Environment Committee, expressed in the proposed amendments
and elaborated in detail in the explanatory statement, the scope of
the regulation needed to be extended to also cover food products
derived from (but not containing) genetically modified organisms.
Fast-track authorisation procedures for products that were 'substantially equivalent' to organic food, as envisioned in the common
position, were rejected (European Parliament 1996a). In sum, the
proposed amendments practically eradicated the main difference
between the regulatory approach of the horizontal safety directive
from 1990 and the changes foreseen with the adoption of the novel
food regulation.

But in contrast to the rapporteur's barely concealed attack on the
very idea of product legislation, dissenting voices in Parliament started
to echo the new economic competitiveness arguments. Some parliamentarians warned that extending the scope of the regulation and
insisting on stricter safety and labelling provisions would stifle the
industries to unacceptable degrees (European Voice, 7 March 1996).
In disregard of this criticism, the following report of the Environment Committee called for substantial changes. All of the proposed
amendments were designed to guarantee more stringent safety provisions and labelling rules, and the report called for extending the
scope of the regulation, as advocated by the rapporteur Roth-Behrendt.
But she had miscalculated the political dynamics. During its 12 March
1996 session, virtually all of the 48 amendments recommended in the
report by the Environment Committee were defeated in the plenary
and only a single amendment passed the floor vote without further

changes. Reiterating the need to face up to the economic implications of ever stricter authorisation procedures and labelling rules, the majority of parliamentarians decided to uphold the general notion of a simplified notification procedure. The majority in the plenary equally disregarded the committee's calls for more extensive labelling provisions than those proposed by the Council common position (European Information Service, 16 March 1996). With Parliament voting predominantly along party lines, the EPP successfully introduced new amendments regarding the type of product information manufacturers would be obliged to provide. Yet despite the fact that Parliament had largely ignored the committee's recommendations and moved further towards an agreement with the Council, a majority of delegates still insisted on changes concerning the mandatory labels. Industry Commissioner Bangemann vigorously defended the much laxer labelling scheme backed by the Council position and referred to the Parliament's remaining hesitations as 'absolute gobbledegook' (European Voice, 14 March 1996) during a speech in the assembly. In September 1996, the Conciliation Committee was convened for the first session.

It is crucial to note that when Parliament voted to reintroduce the amendment that made labelling of genetically modified food over a certain threshold mandatory, the advisory committee of the Creutzfeldt-Jakob Disease (CJD) surveillance unit at Western General Hospital in Edinburgh was yet to reveal the fact that bovine spongiform encephalopathy (BSE) had jumped to humans. That announcement was made on 20 March 1996. The food scare that subsequently spread over Europe must have had an impact on the decisions concerning agricultural biotechnology and, most significantly, may have had influence on votes in the Council later on. But the line of conflict was well established and crucial political decisions were taken prior to the dramatic media coverage that would follow after it had become impossible to conceal the extent to which government handling of the BSE crisis had spun out of control. Some of the compromises over the novel food regulation, on the other hand, were made after the announcement and well within the period of the EU food scare. Both aspects will be taken up for further discussion in Chapter 6. Agreement at conciliation over the novel food regulation was reached only in the third and final meeting of the committee in November 1996, after painstaking work over the details of the labelling rules and some remaining conflicts concerning the scope of the law. Most significantly, the Parliamentarians eventually backed down from their tough stance on labelling. The regulation introduced the criterion of 'substantial equivalence'. The threshold for labelling depended on the level of the presence of genetically modified organisms in the final product and was set at 1 per cent – a level the rapporteur later called 'definitely too high' (European Voice, 28 October 1999) – and the labels would have to include the indistinct warning that the food 'may contain' genetically modified organisms. In the days prior to the final vote in Parliament, where the compromise text adopted in conciliation still needed the approval of the assembly to be passed into law, environmental interest groups lobbied delegates to veto the law (Agence Europe, 10 January 1997 and 21 January 1997). Citing confusing inconsistencies in the labelling schemes between the novel food regulation and other EU biotechnology legislation, the main thrust of the argument advanced by groups including Greenpeace and Friends of the Earth was that the provisions agreed on at conciliation were too ambiguous. The regulation was further considered to be too limited in scope, and it was harshly criticised that the law opened a back door for market approval via a simplified notification procedure that exempted these products from the stricter provisions laid out in the deliberate release directive 90/220. Despite last minute attempts to collect enough votes for a veto, the law was passed on 27 January 1997 (European Parliament and Council of the European Union 1997). Yet alongside the publication of the regulation in the Official Journal, the Commission issued a supplementary statement proclaiming that 'should it appear, in the light of experience, that there are gaps in the system of protection of public health provided for by the existing legal framework, in particular in respect of processing aids, it will formulate appropriate proposals in order to fill those gaps' (Agence Europe, 18 February 1997). In the opinion of an increasingly well-connected set of actors on the European level, the mere existence of product legislation with safety standards only similar to the horizontal biotechnology legislation from 1990 was evidence of exactly such gaps in the area of public health and environmental protection. After the novel food regulation passed, the key to closing this gap was the horizontal directive on deliberate release.

One month after the adoption of the novel food regulation, the Commission announced its plans to reorganise the responsibilities for food health in the wake of the BSE crisis. The renamed Directorate-General for Health and Consumer Protection[3] (SANCO) assumed regulatory oversight (European Commission 1997b). DGs Agriculture and, to a lesser extent, Industry were stripped of parts of their portfolios. The same month, Environment Commissioner Bjerregaard issued a statement announcing her intention to propose new legislative amendments of the original safety directive on deliberate release, several months earlier than initially scheduled.

Summary

Following a forceful and well organised campaign to reframe EU biotechnology policy in terms of economic competitiveness, the advocates of a more accommodating regulatory approach slowly gained ground. The directive guiding the use of genetically modified microorganisms in laboratory research was amended according to the wishes of the primarily affected pharmaceutical industry. The risk assessment procedures included in the law where adjusted and partially relaxed in the light of scientific progress. While the revised legislation retained its basic character, the legislative revision was a success of the biotechnology industries. It came at a crucial time, when the industries reorganised their lobby organisations and merged to form EuropaBio, a single, more diverse interest group, which profited from its new status as the main voice of the industry. But the difficulties in passing the law on novel food, despite the fact that this particular legislation was promoted relentlessly as a cornerstone of the new policy agenda by its advocates, also shed some light on the problems ahead. Most importantly, EU policy dynamics began to change. The struggle with Parliament, in particular in the wake of the assembly's veto of the biotechnology patents bill in 1995, indicated a steep increase in the level of political contestation and the rise of a new policy venue at the supranational level. Parliament not only assumed a formal role in the legislative process. It played this role confidently, intent to gain power and broaden its legitimacy and support-base along the way. Clearly, inter-institutional confrontation was one strategy to gain in profile and consolidate its influence. But in this context, the Parliament did not simply

---

[3]Originally, the DG was known as DG XXIV for Consumer Policy and Consumer Health Protection before it adopted its current, shorter name.

replicate the lines of conflict that already existed inside the Commission and that continued to define the two main coalitions. The emphasis on ethical arguments during the deliberation of the patents bill may have been limited to this particular piece of legislation (and Parliament's opposition indeed faded as soon as core provisions were amended to exempt human cloning from the scope of the law). But the fight over the novel food regulation had already revealed some more subtle shifts in language that would play out more strongly as the political conflict over agricultural biotechnology policy took centre stage. Most noteworthy, in defence of the wide-ranging amendments promoted in the environment committee of the European Parliament, the rapporteur Roth-Behrendt argued that her committee had taken on the issue of biotechnology 'in all the crucial areas where the Council of Ministers failed to deliver a proper common position in the interests of Europe's 370 million consumers' (European Voice, 7 March 1996).

5
The Transformation of the Policy Conflict

The rise of consumer protection

Review of the horizontal safety regulation

Following the extensive deliberation of its new biotechnology policy agenda, the Commission issued a 'Report on the Review of Directive 90/220/EEC in the context of the Commission's communication and the White Paper' (European Commission 1996a), three years after the publication of the Delors White Paper itself. The scope and content of the directive was a major determinant of the Commission's ability to create a more industry-supportive regulatory framework, especially as a result of the so-called 'stop-gap' clause. This article of the directive established that the risk assessment procedure of the directive had relevance for all GMO field releases and their marketing, including those covered under sectoral legislation, because all risk assessment procedures for GMOs in the EU needed to fulfil standards 'similar' to those included in the 1991 directive.[4] For this reason, and more generally because of the directives massive regulatory scope, the reform of directive 90/220 was the linchpin of the Commission's entire biotechnology policy agenda.

The report covered a variety of aspects, including technical issues, ambiguous legal terminology and implementation problems. In its most critical parts, however, referring directly to the objectives stated

---

[4]The interpretation that 'similar' in this context translates into 'without lowering safety standards' is given in the report on the review (European Commission 1996a: 9), but this definition was in no way binding.

in the 1994 follow-up communication, the report set out ways in which the directive can become more flexible in scope and procedures and it indicated how horizontal and product legislation should be reconciled. Specifically, the report identified eight problems that call for legislative reform. In the area of risk assessment, the report concluded that 'insufficient clarification concerning the objectives of risk assessment' had hindered full harmonisation between member states in the area of research and development. Furthermore, the report pointed out the lack of any 'link between administrative procedures and identified risks, which may result in cumbersome procedures for low risk releases' of genetically modified organisms. Concerning administrative procedures, the report identified 'cumbersome administrative procedures and approval system for the placing on the market of products, which have led to delays in approving products' and specifically lamented the 'absence of an active role for the Commission on a number of aspects, including the right to propose simplified procedures'. Finally, with respect to problems of regulatory inflexibility, the report criticised the 'absence of sufficient flexibility for technical adaptation, which prevents regular updating of the Directive to scientific and technical progress'. Such 'updates' were primarily seen as possibilities to establish a slimmer regulatory regime. Based on its own assessment, the Commission committed itself to making full use of its independent powers under the current legal framework to speed up reform, including the adoption of administrative procedures that foresaw reforms of the regulatory regime. But the report on the review also clearly indicated the degree to which substantial reform of biotechnology policy was linked to the reform of the original safety directive on deliberate release. Because of its treaty base, legislative reform of the law would fall under the co-decision procedure and hence involve Parliament as a coequal legislator with the Council. Given the vehement reframing of biotechnology in terms of economic competitiveness from 1991 onwards, co-decision meant that the Commission was entering very uncertain political terrain. Lastly, despite striking a tone generally in line with the new regulatory philosophy of industry-oriented policy revision, the report on the review and the events surrounding its publication could not conceal remaining friction. As the Commissioner in charge, Environment Commissioner Bjerregard issued the accompanying press release (European Commission 1996b). The wording of EU press releases rarely takes the form of partisan policy advocacy, but sometimes the messages convey political signals in less obvious ways. In this case, the Environment Commissioner inserted a final paragraph calling biotechnology 'a fast-moving field that requires constant attention, not least because of its environmental and human health implications. This was confirmed', the Commissioner added, 'by the Commission in its Communication on Biotechnology and the White Paper on Growth, Competitiveness and Employment of 1994'. In reality, the Commission communication of 1994 had of course been the most ambitious formulation to that date of EU biotechnology policy from a virtually exclusive focus on the technology's industrial and agricultural exploitation. As the Environment Commissioner's concluding comments revealed not too subtly to the ever widening circle of actors involved in the issues, the battle over the predominance of underlying policy objectives was not over yet.

Parliament's report on the review of directive 90/220

One year later, Parliament responded to the Commission's report on the review by adopting its own report (European Parliament 1996b). The internal deliberation was uncontroversial and the Environment committee adopted the text with only one vote against, on 2 July 1997, before the report was adopted in plenary session on 15 July 1997. According to the report, Parliament first and foremost promoted the precautionary principle as the precept guiding the revision of EU biotechnology policy. According to this principle, lack of full scientific certainty should not be used as a reason for postponing measures to prevent environmental degradation in case of risk of serious or irreversible damage, Parliament argued. As a result, the report roundly opposed the introduction of simplified procedures such as those introduced under the novel food regulation and insisted instead that the system of case-by-case risk evaluations be maintained and strengthened. The report further supported the introduction of new risk categories that would trigger different levels of assessments as part of the authorisation procedures. Both in terms of the definitions used during risk assessments and in terms of the methodology of risk assessment the report called for a greater centralisation of the procedures at the EU level. Horizontal harmonisation was considered the preferable regulatory approach, and the Parliament opposed the Commission proposals to strengthen sectoral legislation on biotechnology products and specific product-based risk assessment procedures. Instead, the horizontal directive was deemed central to a coherent and transparent legislative framework. The report also introduced the idea to make first time authorisations of genetically modified organisms used in field releases time limited, renewable after ten years based on gathered information and a reassessment of the application. Parliament supported the Commission's goal of strengthening the role of independent scientific bodies in the risk assessment procedure and went beyond Commission recommendations concerning obligatory public consultation prior to field releases, including the right of public access to information.

In addition to these points, raised in direct response to the substance of the Commission report, the Parliament emphasised liability and labelling as areas of key concern. Citing the failure of the Commission to act on the question of potential harm caused by the release of genetically modified organisms authorised under the EU legislative framework, the report called for the immediate incorporation of civil liability clauses into the directive. As regards the question of labelling, the report acknowledged that the Commission considers the issue to some extent, in particular following the adoption of the novel food regulation, but Parliament insisted that labelling provisions be extended. Curiously, the report explicitly stated that the new labelling provisions in the 1997 sectoral law would serve as the starting point for further legislative proposals to amend the horizontal directive. Yet, as the Commission complied with Parliament's call for immediate initiation of the legislative review process, it became clear that some players inside the Commission were prepared to exceed by far the ambiguous standards set by the novel food regulation.

The question of labelling

On the issue of labelling, it appears that the report on the review of directive 90/220 had been subject to some last minute changes. One early press release on the report (European Information Service, 14 December 1996) explicitly highlighted the report's position on labelling and cited its strong wording, according to which 'this question absolutely must be addressed in response to consumers' need for information'. The actual text of the final Commission document was much weaker. In the corresponding section, the report elaborated that it will be essential to address the issue and 'take into account the need to inform consumers and to comply with the international obligations of the Community' (European Commission 1996a: 9). The report also noted that product legislation, not the horizontal safety directive 90/220 under review in the report, was traditionally the place to address labelling issues. This last point was in fact crucial. The original safety directive 90/220 contained practically no labelling provisions. The novel food regulation 258/97, which introduced labelling rules, was limited in scope, since it only covered certain food products. Directive 90/220 hence continued to fulfil the important function of a 'catch-all' legislation that covered biotechnological applications not specifically addressed by sectoral laws, and it did contain applicable and detailed rules on the most relevant aspects: risk assessment and authorisation. Ironically, however, after extensive internal reviews and after both the Commission and Parliament had written reports on the directive in the context of the Delors White Paper, the issue that mattered most in political terms was the very aspect not addressed in that law: the question of labelling. After extensive preparation to revise the horizontal safety law and realign it with the new economic perspective on biotechnology, it was labelling provisions which catapulted directive 90/220 high up on the political agenda. As a result, the economic policy frame of EU biotechnology would finally unravel.

Labelling controversy

In April of 1996 and January of 1997, the Commission passed decisions 96/281 and 97/98, which respectively approved the marketing in the EU of a variant of modified soya developed by the US-based company Monsanto and a strain of maize developed by the Swiss biotechnology firm Ciba-Geigy. While the novel food regulation had been adopted in the first month of 1997, it only entered into force three months later and hence did not yet cover the two products. Instead, approval had still been granted under the horizontal directive 90/220. Since the directive lacked applicable labelling provisions at that stage, the two products entered the market without any specific identification. The patchwork character of the labelling requirements under EU law, and the inconsistencies of the novel food regulation and the deliberate release directive 90/220 in particular, became increasingly difficult to justify under growing scrutiny from the press. When individual EU member states decided to introduce additional labelling provisions in response, this not only led to even more regulatory confusion, but also raised legal issues concerning the free movement of goods under the laws governing the internal market. 'The Commission's most pressing task', one observer noted at the time, 'must be to decide who is in charge of the GMO issue' (European Voice, 29 May 1997). In the light of increasing politicisation, the inter-service Biotechnology Coordination Committee inside the European Commission had lost almost all leverage over the policy formulation process. Its meetings were convened less and less frequently. In addition to the already divisive atmosphere between the different Commission services, Parliament's persistent scrutiny of their respective activities let to an almost complete breakdown of effective internal communication. 'Coordination across services has become simpler', one senior Commission official remarked, 'because it has become impossible' (Interview, European Commission, March 2005).

Labelling and the revision of directive 90/220

As the Commission struggled to find a shared policy outlook, Environment Commissioner Bjerregaard seized the opportunity by linking the question of labelling more and more closely to the planned revision of the 90/220 directive – contrary to the strategy announced in the Commission report on the review of this law, which had explicitly placed labelling in the context of extending product legislation. In February 1997, immediately following the adoption of the novel food regulation, she formally informed the college of the Commissioners of her plans to submit proposals for a complete revision of directive 90/220. She promised a legislative draft by the end of June. In the announcement to the press, the Environment Commissioner's spokesperson singled out the plan to introduce mandatory labelling as a clear signal that the Commission took serious note of the current political and public debate across the EU. Trumpeting the same message, the Commissioner herself made it plain in later statements to the press that 'there is no question of backtracking and adopting less stringent labelling requirements' (European Information Service, 21 June 1997). Time clearly played into the hands of the advocates of a complete overhaul of the EU labelling scheme. Even after the novel food regulation took effect, the approval of agricultural raw products still fell under the horizontal directive 90/220, which lacked labelling provisions. With additional uncertainties concerning processed food remaining, retailers often remained unsure whether the products on their shelves actually contained genetically modified ingredients or not. Consumers, likewise, were often unable to tell the difference. The month the novel food regulation took effect, once lauded as a breakthrough of common-sensual regulation and as a law that finally broke with the tradition of over-regulation and overburdening of the European biotechnology sector, leading food companies attacked the Commission for producing a legal framework that was 'piecemeal and confusing' (European Voice, 1 May 1997). Food companies complained of extra costs resulting from their own attempts to fill the gaps and provide meaningful product information. Unnerved by the inconsistent policy framework, retailers publicly called on the Commission to formulate clear regulatory guidelines.

The Environment Commissioner won a first victory when the Commission (European Commission 1997b) adopted, against fierce internal resistance of the Commissioners for Trade and Industry, a provisional compulsory labelling scheme that amended the existing horizontal directive 90/220 (European Voice, 3 April 1997). The Commission formally adopted the measure as Commission directive 97/35 on 18 June 1997 (European Commission 1997c, see also Agence Europe, 18 June 1997). Ironically, just one year after the internal report on the review of directive 90/220 had called on the Commission to use all available mechanisms for 'internal amendments' to extend the flexibility of the directive in light of the new Commission policy of a more industry-friendly biotechnology policy, the reverse had happened. Making use of its powers to amend the law outside of a full legislative process, the Commission instead extended the regulatory scope of the law. The new approach also went far beyond the labelling scheme of the novel food regulation. Labelling was deemed mandatory both in cases of proven presence of genetically modified organisms in the food ('contains GMO-based substances') and in cases where the presence of such substances could not be positively excluded ('may contain GMO-based substances'). On a voluntary basis, producers who could prove their products to be GMO-free were given the option to indicate so as well. The Commissioner for the Environment hailed the adoption of the Commission directive amending directive 90/220 as a 'the breakthrough in the Commission of the principle of labelling' (European Commission 1997d) and reiterated her intention to propose a full revision of the safety directive before the summer.

A new 'general orientation'

While the Environment Commissioner pressed ahead with plans to use the controversy over labelling as a springboard for a much more ambitious legislative agenda, parallel actions by the Commission President aimed at containing the issue and bringing it back under his control. In February 1997, Commission President Santer had used a speech to Parliament to announce his general, albeit unspecific, support for labelling. 'I am making a plea for the gradual establishment of a proper food policy which gives pride of place to consumer protection and consumer health', he informed the assembly. 'One item I would favour is compulsory labelling in all cases' (Bulletin of the European Union, 18 February 1997). To this end, the Commission President charged the Secretariat-General with the task of drafting a general policy paper. The purpose of the paper was to formulate binding policy principles that would guide the various Commission directorates involved in the upcoming legislative review. Yet even the scope of the reform of the existing regulatory policy appeared to be uncertain at that stage. Just weeks after Environment Commissioner Bjerregaard had successfully pushed for the adoption of a labelling amendment to directive 90/220, Agriculture Commissioner Fischler followed suite and announced plans for similar rules to cover genetically modified animal feed (European Voice, 10 April 1997). Processed animal feed that did not contain viable modified organisms remained unregulated at that time, since the novel food regulation only covered food meant for human consumption. For the first time, the Commission now faced internal calls for regulation that covered the entire food chain. This move offered the chance to end the increasing balkanisation of applicable EU law in biotechnology, but it also highlighted the centrality of the scheduled Commission vote on the general direction of the upcoming reform of the entire legislative framework. As the political weight of the decision increased, the issue was pushed up the hierarchy. Commission President Santer personally took over the task of policy formulation.

Yet when the Santer policy paper on labelling of food products was put to a vote, the President was outvoted by his own Commissioners.

A counter proposal from Agriculture Commissioner Fischler won the approval of the college (European Voice, 29 May 1997). After what was widely perceived by Commission officials involved in the process as a failed attempt to consult with fellow Commissioners, only the Commissioners for Industry, Trade and Foreign Affairs backed the President's proposal. Santer's paper drew on language from the novel food regulation. Fischler, in contrast, had relied heavily on input from DGs Environment and Consumer Protection. His counterproposal went far beyond the provisions of the novel food regulation and Santer's proposal by calling for the complete separation of genetically modified from traditional food and by calling for much stricter labelling rules. As one Commission official commented, the new approach, if adopted, would make the core provisions of the novel food regulation 'obsolete only two months after coming into force' (European Voice, 17 July 1997). Striking an unusually non-conciliatory tone, the chairman of EuropaBio commented that the steps envisioned in the new guidelines were so unrealistically restrictive that their mere consideration would 'threaten the credibility of the Commission' (European Voice, 3 July 1997). Despite strong lobbying efforts by the affected industries, the policy paper, appropriately called the Commission's new 'general orientation' on labelling, was adopted on 23 July 1997 (Agence Europe 25 July 1997). While the paper itself was non-binding, it formulated basic provisions and foresaw a 'clear proactive role for the EU' in the labelling of genetically engineered food 'from the stable to the table'. The new policy guidelines thus covered everything from animal feed, through seed to food products. In line with the strict language of the amendment of directive 90/220, which had earlier been adopted on the initiative of the Environment Commissioner, the new general orientation put the aspect of consumer choice front and centre and called for the clearer indication of whether products consisted of, contained, or were derived from genetically modified organisms. Only a little while afterwards, the Commission released a further statement in late July that specified the guidelines for the general orientation on labelling. The rules reiterated the new approach of covering all products containing or derived from genetically modified organisms throughout the entire food chain and clarified that all products with proven presence of such material would have to be labelled as 'contains GMO material' (Agence Europe, 22 August

1997). The application of the vaguer 'may contain' label, which originally covered a wide variety of products under the rules of the novel food regulation, was restricted to such cases where uncertainty over the nature of the product remained. If the presence of genetically modified material could not be excluded because they were not separated from organic products during processing, shipping and handling, labelling was required. The fact that the rules fell short of dropping the ambiguously worded 'may contain' label completely led to immediate and harsh criticism from the Green group in the European Parliament (Agence Europe, 11 August 1997). Despite some inconsistencies, however, the new approach to labelling practically marked the beginning of the end of the industry-oriented approach to EU biotechnology.

In December 1997, the Commission adopted a proposal for a Council regulation on the labelling of soya and maize. The proponents of the biotechnology industries inside the Commission, who had secured a hard-won victory with the adoption of the novel food regulation only months earlier, were largely locked out of the decision making process and stood by and watched as the regulation was drafted in unusually stringent regulatory language (European Commission 1998a). In its final form, passed as Council regulation 1139/98 (Council of the European Union 1998) in May 1998, the law that replaced the earlier Commission regulation far exceeded the scope and the philosophy of novel food regulation 258/97. The criterion of 'substantial equivalence', introduced in the novel food regulation to exempt products from labelling requirements and speed up product authorisations, was dropped entirely. So were the threshold levels above which labelling was mandatory. After pressure from the Council, labels that included the ambiguous phrasing 'may contain', as allowed under the novel food legislation, were dropped as well. Substantively and procedurally, the text of the Council regulation 1139/98 called the central compromises of the novel food regulation into question. After a wrenching legislative procedure that had spanned almost five years, everything was up for grabs again.

Policy frame transformation

With the entire regulatory safety framework back on the agenda, labelling had turned into the 'watershed issue' (Toke 2004: 153). As the date to reveal the agreed position in the context of the more encompassing draft to revise directive 90/220 drew closer, disagreement inside the Commission continued to slow down the decision making process. Proposals by DG Environment to extend scientific evaluation procedures and grant licenses for genetically modified organisms only for limited periods before they had to undergo further assessments were heavily criticised by the biotechnology industries and subsequently denounced by the Commissioners for Industry and Trade in public statements. Commission officials close to the Trade Commissioner attacked Commissioner Bjerregaard openly for giving in to environmental interest groups and misinformed public opinion (European Voice, 13 November 1997). While the Commissioner for Industry persistently made his dissenting views public, his influence and level of involvement declined steeply. The decision to table a draft for the revision of directive 90/220 at the weekly meeting of Commissioners was eventually postponed in November 1997 due to unresolved differences between the Commissioners for Environment and Foreign Relations, not because of the arguments expressed by the Commissioner for Industry. The looming possibility of a trade war with the US over restrictive policy clauses had become the last mountable argument against the drive for ever-tightening legislation on the marketing of agricultural biotechnology products.

Meanwhile, DG Research futilely attempted to reinstate itself in a leading role by offering itself as the Commission service most familiar with the terrain in scientific terms. In a series of Round Tables with independent scientists, the DG sought to calm the waves of the political debate by claiming to address various aspects of the overall issue in objective terms. One such meeting dealt exclusively with human and environmental risks posed by a single trait of genetically manipulated maize. The experience was less than encouraging. After an entire day of presentations by independent experts, no research team had reported evidence of any risk. Yet when the Commission official in charge of drafting the press release that summarised the results of the discussion left the meeting, he received a call from the Brussels office of Greenpeace. If the favourable assessment of the scientists were made public, the group straightforwardly threatened, it would boycott any upcoming round tables on biotechnology and attempt to sink the Commission's chances at restoring itself as the central EU political actor in the debate (Interview, European Commission, March 2005).

No press release was issued. On a separate occasion, officials from DG Research were told off with unusual bluntness by the head of the Environment Commissioner's cabinet. Confronted by a member of DG Research with the fact that the European Commission's own multi-million Euro research projects had failed to detect any indication of risk resulting from the release of genetically modified organisms under the current rules, the head of cabinet replied that he 'could not care a damn as to what the science says' (Interview, European Commission, March 2005). Nor did he have to as long as the focus of the policy debate shifted in his favour.

Consumer policy coalition

Overt political pressure by environmental groups was still a factor, but towards the end of the 1990s, the political dynamic of biotechnology policy revision had slowly taken on a completely new direction. Public support for genetically modified crops and food was dwindling throughout the period of deliberation over the reform of the deliberate release directive. In the years from 1996 to 1999, support for genetically modified food across Europe dropped from 45 to 36 per cent, while outright opposition rose from 39 to 52 per cent (European Commission 2003a: 16). Throughout the Union, food retailers started to adopt their own labelling policies and worked on the technical possibilities of detecting the presence of genetically modified ingredients at levels as low as 0.1 per cent (European Voice, 28 October 1999). With many of their demands, such as the call to extend labelling provisions to include processed food in which no presence of genetically modified organisms could be detected any longer, retailers were not alone. Instead, their positions were now often identical with the positions pushed for by European consumers and their advocates, primarily the European Consumers' Organisation (BEUC), which serves as an umbrella organisation for the primary national consumer associations of the Unions (see Toke 2004: 184–185). In tandem, the political interests represented by both groups formed a highly convincing alliance. Cooperation among competing retail companies had already emerged over the period when the Commission addressed the issue of labelling and had resulted in coordinated press statements calling for a more consistent policy in this area (European Voice, 1 May 1997). In March 1999, five of Europe's largest supermarket chains finally announced in a joint press conference that they

had founded a consortium to buy foods completely free of genetically modified ingredients in order to market them under their
own labelling scheme (European Voice, 25 March 1999). 'Our policy
is freedom of choice for our consumers', said the secretary-general
of the EU retailer's lobby Euro-Commerce, explaining the position
of his members, 'and that obviously means clear labelling... We
need a lead from the lawmakers on this. We need a clear definition'
(European Voice, 18 March 1999).

As the future of further authorisations of genetically modified
organisms hung in the balance and retailer and consumer took an
increasingly hesitant stance towards food products based on modern
biotechnology, the number of experimental field releases of already
authorised organisms in the EU began a steep decline. Against this
backdrop, the Environment Commissioner slowly changed the focus
on environmental and human protection into an economic argument. Increasingly, the official line or reasoning became that only
the most stringent regulation could revive the confidence of consumers and the public. After years of highly public political conflict,
even the advocates of the economic competitiveness frame started
to concede that deregulation would only fuel public scepticism and
further inhibit the possibility of reaping economic benefits.

Collapse of the economic frame

With the policy shift in focus and emphasis, the political ground
shifted as well. Two dimensions of the debate, economic competitiveness versus human and environmental safety, had shaped the
political conflict over EU biotechnology policy for a decade and a
half. While the two policy frames had been seen as conflicting from
the outset, this logic almost reversed itself. Now, more stringent safety
regulations were the only hope of restoring consumer confidence
and creating a European market for biotechnology products. The
way that this had happened disabled the biotechnology industries
politically. Only one year after Industry Commissioner Bangemann
had fought to avoid mandatory labels for genetically modified food
and pushed through notification procedures that exempted genetically engineered food products from full risk assessments and cumbersome authorisation procedures, the interests he had claimed
to represent now practically rested their case. Recalling a meeting
convened by DG Environment with environmental interests and
biotechnology industries representatives, one Commission official

summarised the position of the industry lobby with the words: 'We will play by your rule book. You just tell us what you want, and we will comply' (Interview, European Commission, March 2005).

In addition to the adverse political environment in which they had to operate, internal frictions plagued the lobby group EuropaBio. By 1998, the interests of the influential pharmaceutical industry had been served through the amending of the contained use directive and the adoption of the patents directive. The questions over legal protection and the problems with indiscriminately burdensome safety regulations in laboratory research had been resolved. Food companies, as a result, had become increasingly isolated (see also van Schendelen 2003: 227). Since pharmaceuticals had been exempted from the scope of the deliberate release directive from the beginning, their main objective now was to retain this status and generally stay clear of the contentious political arena that they had successfully managed to escape so far. As a result, industry representatives refrained from outright attacks on the revision of the regulatory scheme and instead pleaded with the Commission over administrative and legal aspects, such as fast track procedures for scientific risk assessment and further harmonisation of risk categories. Later, EuropaBio increasingly focused on the debate over time-limited field release approvals and the provisions for mandatory reassessments of applications. In light of existing mechanisms that allowed the Commission to consider new information at later stages, the industry group argued that limited approvals, along with already non-discriminatory risk assessments would only lead in the first instance to time- and money-consuming administrative over-regulation. The result, on top of the already extraordinarily high research and development costs and uncertain marketing prospects, would therefore stifle the industry (European Voice, 4 December 1997). On the issue of mandatory labelling, the industry lobby had long given up and officially accepted that new legislation would include new and stricter guidelines. 'The pressure on us', one Commission official involved in the policy process recalls, 'came from the NGOs. And if there had been a bit of pressure from industry, it might have changed things. But there wasn't' (Interview, European Commission, March 2005).

The second White Paper

After a rocky start in the 1980s, biotechnology policy in the EU had initially been a regulatory policy area defined by the concern for human safety and environmental protection. As discussed earlier, this first framing of the issues came under immediate attack, and during the Delors presidency biotechnology policy issues were addressed prominently in the White Paper on Competitiveness, Growth and Employment (European Commission 1993a). In the context of the economic policy outlined in this Commission document, biotechnology policy was portrayed unambiguously as an area of innovative economic choices that heralded great benefits and called for a supportive and flexible regulatory framework. Six years after biotechnology was included in the Delors White Paper, however, the same policy issues were revisited by the Commission. This time, the Commission addressed them in the White Paper on Food Safety (European Commission 2000a). Following an earlier Green Paper on the principles of food law (European Commission 1997a), the White Paper addressed every aspect of food production and marketing 'from farm to table' (European Commission 2000a: 3) and covered issues of agricultural biotechnology ranging from animal feed and seed to genetically modified food. As the cornerstone of this new approach, the Commission announced the establishment of a new independent agency on food safety (see below). Primarily however, the Commission used the White Paper to bundle and advocate over 80 policy initiatives listed in the document's annex and presented them under the new common label of food safety. Shortly before the draft paper was presented, the European consumer cooperation EuroCoop reminded the official chiefly responsible, Consumer Protection Commissioner Byrne, that the time to focus on the creation and smooth functioning of the common market was over and piecemeal policy at the expense of encompassing guarantees of public health and environmental safety was no longer acceptable (European Information Service, 15 December 1999). Heeding their demands, Commissioner Byrne struck while the iron was hot. When he announced the White Paper he summarised the policy proposals it contained as 'the most radical and far-reaching ever presented in the area of food safety' (European Information Service, 15 January 2000). EU biotechnology policy issues, many of which had been contested in the legislative arena for the better part of two decades, were now being integrated in an attempt at transforming EU biotechnology policy 'into a proactive, dynamic, coherent and comprehensive instrument to ensure a *high level of human health and consumer protection*' (European Commission 2000b: 8, original emphasis). Safety and protection, the White Paper established in the very first paragraph, 'must always take priority'. Even the environmental dimension of bio-technology policy along which political alliances had formed and at times transformed the distribution of power across EU institutions, was reduced to a policy consideration that merely played an important role 'in addition' to what the issues were really about: 'to protect and promote the health of the consumer' (European Commission 2000b: 6).

Summary

After advocates of a more industry-oriented approach to biotechno-logy had won a first victory with the adoption of the novel food regulation, the conflict over the future emphasis in EU biotechno-logy policy put the main horizontal safety directive back on the agenda. Yet this time the conflict began to take on a new dimen-sion. The controversy over the labelling of genetically modified food in many ways became the turning point of the debate. At no point during the 20-year period covered here were the internal fragment-ation and the lack of cohesion of the European Commission as obvious and as crucial. The Commission President's attempts to contain the policy dynamics and carefully continue with the envi-sioned reforms were blown out of the water when the majority of his colleagues voted to pursue a new policy agenda that emphasised stricter consumer protection. With trust in agricultural biotechno-logy dwindling across the EU, the new focus would significantly broaden and transform the policy controversy. Seizing the momen-tum, the Commissioners for Environment and Health and Consumer Protection worked to expand the scope of the upcoming reform. The White Paper on Food Safety of 2000 testified to the turnaround of the EU biotechnology agenda since the adoption of the Delors White Paper seven years earlier. As the conflict transformed, the policy dynamics changed in response.

Revising the biotechnology safety framework

Directive 2001/18

Following the row over the question of labelling of food products and the transformation of the biotechnology policy frame along a new dimension of consumer protection, the Environment Commissioner, with only a slight delay, fulfilled her promise to provide the draft of a proposal for a revision of biotechnology safety directive 90/220 (European Commission 1998b). The battle inside the Commission was over. As the principles of the proposal that the Commissioners agreed upon on 26 November 1997 illustrate, the Environment Commissioner successfully persuaded her colleagues early on in the process to fall in line with some of the positions she had long advocated against strong resistance. The agreement on the principles of the proposal 'improves provisions for labelling, introduces the systematic consultation of Scientific Committees, introduces mandatory monitoring of products after their placing on the market, which will be linked to a consent granted for a fixed time period of seven years, increases the transparency of the decision making process and modifies the comitology provisions, introduces new authorisation procedures, confirms the possibility to raise ethical concerns, and clarifies the scope of the Directive' (European Commission 1997e, see also European Commission 1998b: 6–7). The agreement, in other words, addressed and settled virtually all issues that had stirred counterattacks from one or the other side. Testifying to the confidence with which the Commission entered the formal legislative deliberations in one of the most contested and turbulent fields of policy in the EU at that time, the accompanying press statement trumpeted the news of tightened consumer protection as well as enhanced protection of human health and the environment. In the last paragraph of the statement, the Environment Commissioner reiterated the twisted economic rationale of the new regulatory scheme: 'I finally hope that the biotech industry – which potentially has huge growth prospects – will regard these new rules as a clarification and as a basis for building long-term confidence and potential trust with the public' (European Commission 1997e).

The Commission proposal and first reading

Based on the agreement, the proposal for a directive revising the original directive 90/220 on the deliberate release of genetically modified organisms (European Commission 1998b) was adopted on 23 February 1998. The proposal foresaw changes across the entire spectrum of regulatory issues and instruments covered by the original law. In addition to the agreement reached earlier, the proposal specified the common risk assessment principles and clarified remaining questions concerning the directive's scope and definitions. Despite the already far-reaching changes, the Green group in Parliament was quick to discard the Commission proposal as an incoherent mix of deregulation and 'so-called concessions', according to a statement made by four leading members of the party immediately after the adoption of the Commission's proposal (Agence Europe, 26 February 1998).

Parliamentarians pointed out that the proposed changes would have only marginal effects, given that a vast number of food products still fell under sectoral legislation that provided lower standards of human and environmental protection which were inconsistent with the draft directive. In fact, the wording of the proposal had been very cautious on the point of allowing the risk assessment provisions of special sectoral legislation to override the general rules of the directive. The explanatory memorandum accompanying the first Commission proposal stated vaguely that the 'Commission considers that this possibility should be included in the proposal'. Changing the proposal in this respect meant far-reaching consequences not only for legislation immediately linked to the same issues, such as the novel food regulation and pending proposals for similar laws on seeds. More crucially, it would have drawn in the pharmaceutical sector, which remained exempted from other biotechnology legislation in terms of risk assessment and was governed under its own sectoral laws. Parliamentarians further stressed that the draft envisioned numerous clauses and some omissions that, in their view, compromised the objective of encompassing human and environmental protection. They specifically noted the foreseen reduction of the time period during which member states could object to the marketing of genetically modified products under the notification procedure, the failure to impose more comprehensive labelling guidelines, the insufficient attention given to secondary effects of field releases on biodiversity, the lack of provisions on liability, and inadequate rules governing the central register for GMO field releases. Parliamentarians knew they were in a strong position to demand more. Summing up the reservations, MEP Breyer (Green, D) declared that the result of the new directive would be that the 'EU will become a vast field of experimentation that will use the European consumers as a guinea pig' (Agence Europe, 26 February 1998).

Attempts by the Austrian EU presidency in November 1998 to speed up the discussion on the revised directive before the presidency was passed on to Germany were immediately rebutted by the chairman of the European Parliament's Environment Committee. He warned that the Council was headed for inter-institutional conflict if it adopted a political agreement before Parliament had had a chance to discuss the committee's report on the proposed changes, scheduled for early 1999. The Council eventually dropped the issue from its agenda. In January 1999, the Environment Committee adopted its report by 16 votes to zero. Eleven abstentions, however, combined extreme positions for either more or less stringent regulation. Yet at this stage, none of those MEPs in favour of a more industry-oriented approach was willing to voice his or her opposition more forcefully than by abstention. The report, drafted by David Bowe (PES, UK), reflected a compromise position based on 188 amendments tabled by members of the committee. It was adopted by the assembly with only a few changes (European Parliament 1999). One central parliamentary amendment, subsequently adopted by the Commission in its modified proposal of 25 March 1999 (European Commission 1999a), concerned a more explicit reference to the precautionary principle. Several of the other amendments, however, were decidedly more far-reaching. Parliament called for the *de facto* extension of the risk assessment laid down in the revised directive to all vertical product-based legislation. To this end, it introduced amendments that made the revised directive the standard against which all sector-specific laws would have to be judged as 'at least equivalent' in terms of their risk assessment procedure. Parliament further addressed the issue of traceability, demanding that records for all products placed on the market must be kept so as to facilitate their complete recall in warranted cases, and it called for more definitive labelling rules, denouncing the 'may contain' clause still included in the draft. In what probably amounted to the most severe blow to the biotechnology industries, the parliamentary report also called for full civil and criminal liability for any damage to human health or the environment resulting from the deliberate release of genetically modified organisms, in the absence of a more encompassing EU environment liability laws. As a surprise to many observers however, the report by Parliament also proposed to extend the time limit of initial field releases from the proposed seven to 12 years, arguing that the Commission's position was unnecessarily restrictive. Pushed forward by a Parliament bristling with self-esteem and under intense political pressure from the Council, the responsible DG

Environment began work on further tightening the directive and seeking ways to expand the scope of the regulatory framework. 'There was no competing with that from the other DGs that might have been disposed to fight it', a Commission official recalled, 'Research, Industry and Agriculture and Trade were essentially silent' (Interview, European Commission, March 2005). The precautionary principle entered the revised Commission proposal (European Commission 1999a), albeit less forcefully than the Parliament had intended. In following the Parliament's demands, however, the Commission adopted forty amendments partially or in their entirety, including the extension of the risk assessment provisions of the horizontal directive as a binding standard for product legislation. The adoption of this amendment was a significant step towards tightening and harmonising the entire EU biotechnology regulation scheme. The Commission, however, rejected the majority of these substantial amendments by Parliament. The inclusion of secondary and long-term effects on the environment under the environmental risk assessment, questions of limits on first time approvals for experimental releases, issues concerning traceability, labelling as well as civil and criminal liability, remained unsettled when the Commission forwarded its modified proposal to the Council.

The de facto moratorium

The Council was originally scheduled to address the proposal at its 11 March 1999 meeting of environmental ministers, but due to wide-ranging disagreements among the member states and delays in the Commission's response to the amendments made by Parliament, the Council did not discuss the directive until its June meeting. In preparation of the meeting, the German presidency reverted to what was widely perceived as a political escape strategy by including in its report one of the most contentious amendments introduced by the Environment Committee in Parliament. According to the report tabled before the members of the Environment Council, all future approvals for the release of genetically modified organisms were to be granted by an independent agency.

Clearly, the establishment of an agency reduced the likelihood of prolonged political battles between the EU institutions that were responsible for approvals under the original directive or under the Commission proposal for revised legislation. At the same time, however, the delegation of the task of granting approvals to an independent regulatory agency appeared to conflict with the overwhelming call for more transparent and democratically accountable decision making procedures in the EU biotechnology sector. 'They have deliberately put forward a report that will not be accepted', the spokesperson for Commissioner Bjerregaard protested, pointing out that the plan for an independent agency 'does not stand a snowball's chance in hell' (European Voice, 3 June 1999). Details of the draft directive were discussed in Council working groups. All proposed changes substantially extended the safety provisions included in the Commission text, but a common position was not formally adopted and a decision was thus effectively postponed. During its 24–25 June 1999 meeting however, the Environment Council reached agreement on the dramatic step of declaring a *de facto* ban on all approvals of genetically modified releases and their marketing until a new encompassing biotechnology framework had been adopted. This unusual step only added to the already crushing political pressure on the Commission to resolve the issue conclusively. Approvals of genetically modified organisms under directive 90/220 had not been given since October of 1998, and thirteen applications were pending at the time of the meeting (Christoforou 2004: 691). As it remained unclear whether there existed a legal basis for a complete suspension of these approvals, the statement jointly adopted by the Council and the Commission only justified the postponement of a decision on the directive under discussion. The statement pointed out that Ireland was presently in the middle of a national consultation process leading up to a planned referendum, and it was argued that the outcome of this referendum needed to be taken into account before a definitive position could be taken. Eager to send a stronger signal, France subsequently tabled a second declaration backed by the representatives from Denmark, Greece, Italy and Luxembourg. These countries called upon the Commission to submit 'without delay full draft rules ensuring labelling and traceability of GMOs and GMO-derived products and state that, pending the adoption of such rules, in accordance with preventive and precautionary principles, they will take steps to have any new authorisations for growing and placing on the market suspended' (Council of the European Union 1999). Choosing a somewhat less definitive wording, Austria, Belgium, Finland, Germany, the Netherlands, Spain and Sweden adopted yet another declaration of their intent not to authorise any new release at the present stage. Equally stressing the urgent need to take a more precautious approach towards the release of genetically modified organisms, they called for immediate adoption of traceability and labelling provisions under the Commission's executive powers. This second declaration also evoked the right to enact accompanying measures to this effect within the EU biotechnology safety framework at the respective national levels in case the Commission should fail to comply with their priorities. Only the United Kingdom withheld its support for the dramatic step.

The declaration of the *de facto* moratorium by the Council did not come as a complete surprise, nor was the Council the first European institution to consider this strategy. The Council's declarations had been preceded by a letter of the Parliament's Environment Committee chairman Ken Collins (PES, UK) to the Commission President in which he called for a stop of all approvals and asked the Commission not to start infringement procedures against those member states already blocking new approvals under Article 16 of directive 90/220. And Parliament was not alone. Ever since 1997, DG Environment had been accused of exploiting its formal role in the approval process by failing to pass on its assessments of application dossiers to the regulatory committee, thus practically freezing the process of field releases and market approvals (European Voice, 6 November 1997). During the meeting of Environmental ministers in July 1999 when the *de facto* moratorium was adopted, the Environment Commissioner had repeatedly alluded to the possibility of such a step (European Information Service, 26 June 1999). What appeared to be a confrontational step of intra-institutional politics was thus in fact only the final, albeit decisive, step by a coalition among political institutions of the Union towards building up political pressure directed at radical policy reform. In response to the declaration of the *de facto* moratorium, the Commission issued a statement on behalf of Commissioner Bjerregaard. It concluded by noting that the Commission 'recognises that the declaration put forward by France also underlines the importance that Member States attach to the Commission presenting without delay rules for ensuring labelling and tractability of GMOs and derived products'. By way of helping the Commission to adopt legislative proposals to this effect, member states were invited to communicate their policy objectives to the Commission immediately. The most pressing issue on which the Commission sought prompt input from the national governments, the last sentence of the statement emphasised, was the question of labelling (European Commission 1999b).

Since a formal moratorium of the release and marketing of genetically modified food and feed was in conflict with existing EU law,
a legally binding decision to this effect was never taken. While the
right to grant approvals under directive 90/220 lay primarily with
the Commission, member states could block the implementation of
the decisions under a safeguard clause in directive 90/220 and, following the 1999 meeting of the Environment Ministers, DG Environment compliantly went along and no new approvals were granted.
When the presidency passed on to the French government, under
pressure to protect traditional farming interests, it was clear that the
de facto moratorium on the approval of genetically modified organisms would remain in place until the issues had been addressed
to member states' satisfaction (European Voice, 3 August 2000). The
ban was widely seen as a concession to environmental groups and
the Parliament, but it clearly also reflected mounting pressure on
national governments to take a tougher stance on issues of food
safety. In particular, a ruling by the French *Conseil d'Etat* to ask for a
European Court of Justice (ECJ) opinion on the issue of three French
authorisations of genetically modified maize strains earlier that year
had helped proponents of stringent regulations gain ground on the
member states level. Greenpeace France had taken the issue before
the French courts, arguing that approvals had not followed proper
procedure, and as a result of the ruling of France's highest constitutional court, the sale of the products had remained prohibited until
the ECJ ruling was handed down. Notwithstanding the open legal
issues in France, however, the Secretary General of DG Environment
stated the obvious with regard to the *de facto* moratorium: 'At the
end of the day, the decision is a political one' (European Voice,
17 December 1998).

Common position and second reading

During the same meeting on 24–25 June 1999 when the moratorium
was called, the Council reached political agreement on the principles
of its position concerning the revised Commission proposal, and
in December 1999 the Council finally adopted a common position
unanimously. With one exception, the Council text adopted all the
changes made by Parliament that had also entered the modified Commission proposal. In addition, the Council included a substantial
number of those Parliament's amendments which the Commission

had previously dropped, including a more forceful application of the precautionary principle, a ten-year time limit on authorisation, and the insistence on rules ensuring traceability, which pertained to the origin of genetically modified organisms placed on the market. While Parliament was waiting to receive the Council position, the political balance in the institutions shifted. The European People's Party (EPP) ousted the Party of European Socialists (PES) as the strongest political group in the European Parliament elections, reviving hopes that the assembly would reconvene with a more industry-leaning majority and reconsider its position on the issues. Prior to Parliament's receipt of the Council positions, EPP members had still maintained that concerns for maximum safety had to be weighed against negative effects on industry and also had to be re-evaluated in the light of Parliament's role in the *de facto* moratorium (European Voice, 6 January 2000). EuropaBio lobbied Members of the European Parliament ahead of a crunch vote in the Environment Committee, but the best the biotechnology industries lobbyists could now hope for was that Parliament would go into conciliation. This step would have given the biotechnology industries more time and allowed their lobby to explore its chances to shift the balance on some issues among the new, more conservative deputies (European Voice, 10 February 2000 and 13 March 2000). In second reading in April 2000, the Parliament in fact amended the recommendations from the Environment Committee in the plenum before adopting the resolution on the Council common position. But while changes made by the assembly dew predictable criticism from some Green group and socialist members, the effects on the regulatory framework were marginal. The adopted position backed away from the assembly's earlier demands to include liability clauses in the revised directive, but only after the new Environment Commissioner Wallström promised to introduce legislation on liability before the end of the following year. Parliament further dropped an amendment detailing requirements aimed at preventing the so-called transboundary movement of genetically modified organisms, a requirement that biotechnology industries and independent experts had deemed impossible to implement in the first place, in exchange for tougher assessment criteria based on this aspect. In following the committee recommendations, the plenary reintroduced the compulsory assessment of accumulated long-term effects of field releases on the environment and agreed to the ten-year time limit on authorisations proposed by the Council (European Information Service, 15 April 2000).

Among the few critics of the overall position, the Green MEP Paul Lannoy and the Green Environment Minister of France, Dominique Voynet, stood out. The exclusions of liability clauses and of the regulation of transboundary movements were at the core of their criticism, but even environmental organisations like Greenpeace and Friends of the Earth struck mild tones and largely acknowledged that progress was being made even from their viewpoint. Otherwise, reactions to the vote in Parliament were positive across the board. Commissioner Wallström welcomed the compromises almost unambiguously. Joining in the praise, EuropaBio issued a statement in which it 'welcomes today's vote by the European Parliament of the revision of EU-Directive 90/220'. The Commissioner lauded the 'balanced outcome' as a right step taken towards a biotechnology framework that will 'contribute to a more stable investment climate, improved competitiveness and additional jobs'. Most of all, the statement notes, the vote pave the way for the adoption of the directive and hence promised 'clarity, predictability and certainty' for the European biotechnology industries (Agence Europe 26 April 2000). In fact, the level of certainty reached at this stage was already sufficient for one of the multi-national food manufacturer to shut down parts of its business in Europe. Novartis, in a public letter pointedly addressed to Greenpeace of 2 August 2000, announced that all products manufactured by its suppliers were free of genetically modified organisms. The company continued, however, to produce the modified maize that had been at the centre of the political debate throughout the mid-1990s (Agence Europe, 9 August 2000).

Agreement at conciliation

The Commission (European Commission 2000b) accepted 13 parliamentary amendments fully or in principle, while dropping 16 others, often on legal and technical grounds. The substance of the changes, however, was reflected in the Commission opinion. On 15 September 2000, the Council of Environmental Ministers initiated the conciliation procedure by declaring that it could not accept the changes Parliament had proposed in its second reading in their entirety (see European Information Service 13 September 2000). A first informal trialogue meeting between the Commission and both legislative bodies was convened, and immediately following the first informal meeting on 19 October 2000, it became clear that an agreement on several issues was well within reach. Among the issues still considered contentious was the inclusion of pharmaceuticals. The question whether or not, and if so to what extent, pharmaceuticals should fall under the horizontal directive or remain exempted had split the Parliament. Marketing of pharmaceuticals would remain governed by special product legislation, according to all participating parties in the trialogue. Questions remained, however, whether pharmaceuticals based on genetically modified organisms should be exempted from the risk assessment under the horizontal directive at the research and development stage. Parliament had adopted two contradictory amendments on the issue, but the representatives signalled early in the meetings their willingness to concede to a total exclusion of pharmaceuticals. Still under discussion was the possibility of a simplified procedure for approvals. Parliament had reintroduced the measure for cases considered low risk after the Council had taken a more restrictive stance in its common position. On the other hand, Council was proposing a less rigid regulation of renewed approvals. The ten-year limit for first-time authorisations was a compromise all sides could accept, but Parliament insisted that after that period, renewed approvals should be equally limited to another ten years, contrary to Council's view (see European Information Service 31 October 2000). Proving to be the most difficult issue in conciliation was the Parliament amendment calling for a national register that indicated the locations of field releases in each member state. Both the general notion of such a register, as well as the administrative issues involved and the question of access rights had led Council to resist the proposed amendment. After the wording had been changed to allow for substantial leeway in terms of its implementation, the Council accepted mandatory registers. The meeting on 6 December 2000 settled the remaining disputes as well. Parliament succeeded on the issue of renewed market authorisations, which remained limited to a maximum of ten years. Pharmaceuticals would remain covered by product legislation that included safety precautions equivalent to those in the horizontal directive. In principle, however, pharmaceuticals had managed to escape. On 20 December 2000, the Council acknowledged the receipt of a formal letter from Parliament agreeing to the changes made at conciliation (European Information Service, 13 December 2000 and 20 December 2000).

Revised safety directive 2001/18

Parliament and Council voted to adopt the changes made at conciliation on 14 and 15 February 2001 respectively, and signed the text into law on 12 March 2001 (European Parliament and Council 2001). The plenary vote in Parliament supported the compromise reached at conciliation with 338 to 52 votes and 85 abstentions. There were no votes against the compromise in the Council, and only France and Italy abstained. Directive 2001/18 entered into force on 14 April 2001 and as of 17 October 2002, repealed the original safety directive 90/220. In comparison to its predecessor and judged against the current legislative standard in the EU, one expert from the Commission's Legal Service observed that the safety regulation included in the law 'can be considered to have achieved a level of harmonization that is nearly complete' (Christoforou 2004: 671). At the core of the directive lies the environmental risk assessment procedure. It overcomes many of the ambiguities of the original law, which lacked clear principles and was notorious for the inconsistency with which member states interpreted and executed it at the national levels. Directive 2001/18 instead includes the precise objectives, principles and procedures for risk assessment in the annexes of the law. While this serves to make the risk assessment more coherent, the law also significantly extends the scope of the procedure by including more than the direct and immediate effects of the genetically modified organisms on their environment. Risk assessment under the new directive further includes cumulative long-term effects on human health and the environment (see annex II), as well as an assessment of possible indirect and long-term effects (see also Lawrence et al. 2002). Approvals for the field release or marketing of a genetically modified organism are obtained first from the competent national authority. In this respect, directive 2001/18 constitutes a hybrid between two types of EU regulatory approaches, with the mere harmonisation of national procedures as one extreme and the complete centralisation of decision making at the Community level as the other (Christoforou 2004: 678). Following the criteria and processes detailed in the directive, the competent national authority assesses the application and grants or rejects the right for experimental release in its own territory. All member states' competent authorities become involved when the application is for the marketing of a genetically modified organism across the EU. In such cases, the national authority in the country where the application was made forwards the dossier to the Commission, which then forwards it to the member states. Like the original horizontal legislation, directive 2001/18 includes a safeguard clause that allows Member States to restrict or revoke the authorisations of genetically modified organisms in their territory. This right, listed in Article 23 of the new directive, is significantly more limited than it was under the previous regulatory scheme. It requires in its current form that a detailed case be made for a risk to human health or the environment that was not taken into consideration during the assessment of the dossier (see also Tsioumani 2004: 285–286; Lawrence et al. 2002: 51–52). Time limits for first authorisation were set at ten years. The public gained access to field release registers under the new law, which also made it obligatory to consult with the public and interest groups on any decision to authorise the release of genetically modified organisms. Mandatory consultations at the EU level include with scientific expert committees and the European Parliament. The decision to adopt or reject a Commission proposal for authorisation under directive 2001/18 lies with the Council, which acts by qualified majority. A possibility to introduce a simplified procedure in cases where experience with product safety is deemed sufficient was not included in the final version of the law.

In addition to the more stringent authorisation procedures, directive 2001/18 includes mandatory post-marketing monitoring requirements that enable the authorities to 'trace and identify any direct or indirect, immediate, delayed or unforeseen effects on human health or the environment', as the corresponding articles of the law explain. Monitoring plans are an integral part of the overall application, and consideration of the cumulative long-term effects of genetically modified organisms on the environment became part of the risk assessment procedure prior to authorisation. While the old directive 90/220 is considered an early example of precautionary policy making (e.g. MacKenzie and Francescon 2000: 533), it was written and adopted well before the term came into common use and began to stir criticism from advocates of a less risk averse regulatory philosophy. The new directive adopts the precautionary regulatory approach explicitly in its preamble and then refers to it repeatedly throughout the text of the law, both in general terms and in the context of risk assessment (see also Francescon 2001). Member states were given greater say in the approval process, but as noted above, their rights to temporarily ban products authorised under this directive from their markets were curtailed. With the adoption of the law, the Commission committed itself in written declaration to bring forward further legislation before the end of 2001 on environmental liability, covering possible damages resulting from released genetically modified organisms, and on traceability and labelling. The latter law was to specify provisions on labelling already included in directive 2001/18, which generally states that labels are required at all stages of the placing on the market. EuropaBio, the biotechnology industries association, conceded defeat: It 'welcomes the positive signal the European Parliament has given in approving the revision of Directive 90/220 on the deliberate release of genetically modified organisms... [The] Directive at last leads the way to establishing a more rigorous and coherent framework for the regulation and market supervision of biotechnology in Europe.' Claiming that the 'European biotechnology industries association has always advocated rigorous regulations', the statement continued, 'EuropaBio believes the amended Directive will further strengthen the already stringent safety assessment process, help to establish consumer confidence in the regulatory process and convince investors that there is a future for agro-food biotechnology in Europe' (EuropaBio 2001).

Extending biotechnology safety regulations

Like its predecessor, directive 2001/18 includes exemption clauses from the risk assessment procedures for pharmaceuticals, which remained covered under special sectoral legislation. Equally, directive 2001/18 did not cover products derived from or produced with genetically modified organisms if they are no longer present in the final product. But with the adoption of the new horizontal safety directive 2001/18, the political struggle over the reform of EU biotechnology policy was not entirely over. Labelling and traceability provisions in the new directive were kept general, and Parliament only approved the law after the Commission had promised to propose follow-up legislation to address these issues immediately. Similarly, the new safety framework needed to be amended to include food so far covered by the less stringent novel food regulation and animal feed that remained unregulated at that time. The risk assessment, authorisation and monitoring of genetically modified food and feed, as well as their labelling, were addressed by the Commission proposal for a regulation (European Commission 2001a) of 25 July 2001. The traceability and labelling of genetically modified organisms covered by directive 2001/18 was addressed by a second proposal for a regulation (European Commission 2001b) on the same day.

Food and feed

The specific application of the risk assessment, authorisation and monitoring rules of the food and feed regulation followed the general provisions of the new horizontal safety directive 2001/18, which specifies that sectoral legislation may only override its applicability if the standards are 'at least equivalent' to those of the directive 2001/18. The Commission proposal (European Commission 2001a) foresaw the implementation of the so-called 'one door, one key' approach, according to which risk assessment and approval procedures were centralised and simplified. The newly created European Food Safety Authority was charged with the task of conducting risk assessments based on which the Commission proposed to grant or deny authorisation. A member state committee would take the final decision, granting authorisations limited to ten years initially, subject to renewal. In comparison with the novel food regulation of 1997, which continued to cover traditional food but was replaced to the extent that it covered genetically modified food products, the proposed risk assessment and labelling rules were naturally much stricter. Most importantly, the criterion of 'substantial equivalence', introduced in the novel food regulation, was abandoned. Under the old regulation 258/97 products classified as substantially equivalent to an organic or conventional counterpart had been exempted from mandatory labelling and full risk assessment and authorisation procedures. While falling short of scrapping the criterion of 'substantial equivalence' entirely, the text of the new regulation 1829/2003 explicitly breaks with the philosophy of the earlier law and considers substantial equivalence with organic food products 'not a safety assessment in itself' (recital 6). As a result, the simplified notification procedure that made a full authorisation procedure for 'substantially equivalent' products unnecessary under the old regulatory scheme was eliminated. Furthermore, the scope of the regulation was significantly extended to cover food products produced from genetically modified organisms, irrespective of whether or not traces of the organisms were actually present in the end product, as well as animal feed, which had remained unregulated up until then (see also Francescon 2001: 317–319).

The central issue of political debate proved to be the question of labelling thresholds. According to the Commission proposal, labelling was compulsory for all food or feed containing, consisting of or produced from genetically modified organisms. However, the draft included a derogation clause according to which labelling rules did not apply to organic food or feed products that tested at one per cent or lower for genetically modified organisms. Such a threshold level takes account off the so-called adventitious presence of genetically modified organisms that can result from natural cross-fertilisation, or can occur at any stage of harvesting, storing, shipping and processing of food or feed unless it is kept completely separate from genetically modified production and processing equipment (see European Information Service 27 July 2001). In its first reading of the proposal on 4 June 2002, EU parliamentarians adopted 111 amendments. Despite the high number, the issues of substantial disagreement with the Commission were in fact limited. The assembly voted to change the threshold level under which labelling rules should not apply to 0.5 per cent. Moreover, the amendments by Parliament scrap all derogation clauses for genetically modified organisms entering the market, that were not authorised under EU law, insisting that their exemption would undermine the entire biotechnology safety framework. An amendment that extended the labelling provisions to food derived from genetically modified animals feed had been included in the report of the environment committee, but was not adopted by the assembly. Based on the argument that the present text of the regulation would give rise to unfair trade conditions, the Parliament further insisted that all provisions of the law apply also to products imported into the EU. In a clear affront to the Commission, Parliament determined that the proposed regulation did not constitute a sectoral law in the meaning of Article 12 of directive 2001/18. Challenging this status meant that risk assessment and authorisation procedures could not follow the 'one door, one key' policy, but would instead be mandatory under both legislations. The bulk of the other changes introduced by Parliament strengthened public access to information and extended public involvement in the consultation process under the law. Along with almost half of the amendments presented by Parliament, the Commission's amended proposal (European Commission 2002) did take up the extension of the scope of the regulation to cover imports into the Union. It did not include, however, the changes aiming at suppressing the derogation for adventitious presence of unauthorised genetically modified organisms, nor did the Commission accept the Parliament's amendment denying the regulation the status of sectoral legislation. On the question of labelling thresholds, the Commission opted for a double strategy. It argued that thresholds of 1 per cent offered sufficient protection, but excluded any specification in its revised proposal. Instead, threshold levels were to be determined by comitology procedure. Based on the revised proposal, the Council on 17 March 2003 adopted a common position that struck a balance between the diverting views of the Parliament and the Commission on most of the remaining critical question. Derogation from authorisation was set at a level of 0.5 per cent presence of genetically modified organisms. Labelling thresholds were set at 0.9 per cent and subject to revision through comitology procedure. The role of member states *vis-à-vis* the new European Food Safety Authority was strengthened. In a blow to the Commission policy of 'one door, one key', the Council followed Parliament in limiting the extent to which risk assessment and authorisation under the planned regulation exempted applicants to seek further approval under the horizontal directive 2001/18. Parliament proposed only nine further changes, all perceived by the Commission as compromise solutions, and Parliament and Council had agreed on the final version by September 2003. As adopted, regulation 1829/2003 (European Parliament and Council 2003a) on genetically modified food and feed repealed in its entirety Council Regulation 1139/98 as well as Commission Regulations 49/2000 and 50/2000 on the labelling of certain foodstuffs, and it amended the core provisions of the regulation 258/97 on novel food, thereby dramatically limited its applicability and scope (see also Tsioumani 2004).

Traceability and labelling

Under the regulation on traceability and labelling, the safety framework was further extended, building on the concept of traceability first introduced into EU legislation in the horizontal directive 2001/18. As outlined in the explanatory memorandum of the Commission proposal (European Commission 2001b), traceability aims at a system of identifying products for the purposes of controlling labelling rules, monitoring potential effects of authorised products, and facilitating their recall in the event of unforeseen risks to human health or the environment. As adopted, regulation 1830/2003 on traceability and labelling of genetically modified organisms (European Commission 2003b) thus amended directive 2001/18 and served to achieve some of the objectives of regulation 1829/2003. It regulates that information on products consisting of, containing or produced from genetically modified organisms must be retained at each stage of the process of marketing. It requires that labels indicate the genetically modified organisms contained in a food or feed, or the organisms from which the food or feed has been produced, over the threshold of 0.9 per cent and 0.5 per cent for accidental contamination with genetically modified organisms that are unauthorised in the EU. The regulation thus extended the EU labelling regime to cover animal feed and set requirements for both food and feed irrespective of whether genetically modified organisms are in fact still present in the product. Furthermore, all scientific risk assessment procedures were ruled to be carried out by the European Food Safety Authority, whose opinions was to be made public and accompanied by a public register of all authorised products. In 2003, following the written declarations from the Commission to Parliament, the two follow-up laws were adopted by Parliament and Council (European Parliament and Council 2003a, 2003b), completing the new regulatory framework at the centre of which lies directive 2001/18. The day the two regulations 1829/2003 and 1830/2003 were adopted, Consumer Protection Commissioner Byrne issued a joined statement with Environment Commissioner Wallström. 'European consumers can now have confidence', he was quoted in the Commission press release, 'that any GM food or feed marketed in Europe has been subject to the most rigorous pre-marketing assessment in the world' (European Commission 2003c).

European Food Safety Authority

In the context of the adoption of the food and feed regulation, the European Food Safety Authority (EFSA) was established by a European Parliament and Council regulation of 28 January 2002 (European Parliament and Council 2002). The idea of a regulatory agency in the area of food was first advanced in the White Paper on Food Safety in early 2000. Originally based in Brussels, EFSA was given a management board by the end of 2002 and became fully operational as a European agency by May 2003 with the establishment of its various scientific boards. The agency was later moved to its permanent location in Parma. As a legal entity separate from the European Commission and other EU or national institutions, EFSA and its director are only answerable to the agency's own management board. Unlike independent regulatory agencies in other political contexts however, EFSA was not delegated any decision making authority (see Majone 2002: 390). The agency's primary task is to provide and coordinate scientific risk assessments under the various food safety legislations in the Union, including the revised biotechnology safety framework, and collect and analyse data towards this end. Importantly, these recommendations include the scientific evaluations based on which the Commission drafts the proposals for the granting or refusal of authorisations under the reformed biotechnology regulations. Yet EFSA's recommendations are non-binding and only provide scientific advice to the Commission, which 'remains fully responsible for communicating risk management measures' under the various authorisation procedures, as stated in the recitals of regulation 178/2002 by which the agency was established. The regulation further created the Standing Committee on the Food Chain and Animal Health (Article 58), a body chaired by the Commission, but composed of over 300 experts delegated as national representatives, that supervised the Commission's implementing decisions under the comitology system.

Breakdown of policy implementation

Since the regulatory overhaul between 2001 and 2003, the EU legislative agenda in biotechnology finally reached a steady state. The few amendments and extensions of the framework that were adopted in the following years remained technical in nature and predominantly pertained to the implementation of the legislation in place. Among these changes, Commission regulation 65/2004 (European Commission 2004a) of 14 January 2004 created new obligations concerning the traceability of GMOs placed on the market by way of establishing 'unique identifiers', each made up of a combination of numbers and letters, which were included in the respective product label. This system extended to the EU a format developed by the Organisation of Economic Cooperation and Development (OECD) and linked to the so-called BioTrack product database run by the OECD department on biosafety. Regulation 1829/2003 on food and feed was further clarified by Commission regulation 641/2004 (European Commission 2004b) of 6 April 2004 as regards the rules governing product notifications, and additional guidelines were established for the transition to the new regulatory regime established by the 2003 law for applications and notifications issued prior to the date the new rules came into force. With respect to the horizontal safety directive 2001/18, Commission decision 2004/204 (European Commission 2004c) adopted changes to the operation of the central registers of genetic modifications. It furthermore became necessary to adopt new rules on EU committee oversight after a more general reform of the comitology system. In 2006, the system of implementing powers conferred on the Commission was extended to include a 'regulatory procedure with scrutiny' for implementing decisions that affect the general scope of the law. By adopting directive 2008/27 the Parliament and Council (European Parliament and Council of the European Union 2008) amended safety directive 2001/18 in accordance with the new procedures, allowing Parliament and the Council to block Commission measures if they affect central part of the law's annexes, criteria for notification and threshold levels.

While legislative reforms remained limited, problems with the implementation of the new regulatory regime became ever more evident. In January of 2004, following an internal debate on legislative progress and regulatory implementation, the Commission issued a joint statement of the Commission President and the six Commissioners involved in issues of biotechnology (European Commission 2004d). The key message of the communication was that the 'the new regulatory framework for GMOs has been completed', but member states failed to live up to their treaty obligations and executive functions under European law. Despite the rigorous rules and regulations, seven counties still invoked the 'safeguard clause' and banned genetically modified products long authorised in the EU from their territory. Equally troubling, the intricate authorisation process of new genetically modified products stalled when put to the test for the first time. The Standing Committee on the Food Chain and Animal Health, which issued opinions on new applications based on the scientific dossier supplied by the European Food Safety Authority, did not reached a sufficient majority to take a decision on the pending authorisation of genetically modified sweet maize Bt11. The dossier was forwarded to the Council. The Council, however, did not reach a sufficient majority either. Neither approved nor rejected, the decision fell to the Commission, which used its powers to issue the approval – the first in six years (European Commission 2004e).

Over the following years, the initial problems with the implementation of the new regulatory framework revealed themselves as a highly problematic pattern. The authorisation procedure remained dysfunctional as member states, both through their representation in the regulatory committee and in the Council, persistently remained unable to summon the sufficient number of votes to decide upon pending applications. Bans on authorised products in five member states (Austria, France, Germany, Greece and Luxemburg) remained in place after the Council voted in June of 2005 to uphold the restrictions, contrary to a Commission request and despite the fact that EFSA rulings did not find scientific grounds on which to sustain the bans, as was required under the safeguard clause in article 23 of directive 2001/18 (European Commission 2005a). By 2006, France had still not transposed the new directive on deliberate release and, following rulings of the European Court of Justice and two written warnings, the Commission finally asked the Court to fine France in excess of €38 million plus a daily penalty. The following year, the Commission (European Commission 2007a) published its second report on the decisions taken under the new safety directive 2001/18, covering the first three years since the law entered into force. Thirteen applications for genetically modified plants had been processed. Of these 13, five products had been authorised. In all five cases included in the Commission's 2007 review, as well as in the cases of the four additional authorisations that were granted until the summer of 2010, the responsible regulatory committee and the Council failed to adopt a position by qualified majority every time and thereby left the authorisation decision to the Commission (European Commission 2004e, 2005b, 2005c, 2005d, 2006, 2007b, 2007c, 2009, 2010a).

The persistent stalemate over authorisations granted under the new regulatory framework finally led the Commission to revise its strategy. In 2010, the same day that the Commission adopted decisions on five authorisations and cleared genetically modified maize and potatoes for marketing or use in industrial processes and animal feed, the Commission announced that it would seek a revision of the authorisation procedure in place. The new procedure, the Commission announced, would need to maintain its grounding in scientific opinions, but would aim at giving member states more choice over whether or not to cultivate genetically modified plants (European Commission 2010b). The Commission adopted the view, also voiced previously by the Dutch delegation to the Standing Committee on the Food Chain and Animal Health during a February 2009 meeting, that such revisions would not have to conflict with the rules of the internal market. Products imported and authorised under existing regulations would still be marketed across the EU, whereas the decision of whether or not to allow cultivation of genetically modified plants would be left to the individually member states. In essence, however, the Commission was plainly seeking out provisions that would establish legal grounds for member states' failure to implement EU biotechnology law.

The Commission intentions were reiterated in a communication of 13 July 2010 (European Commission 2010c), the same day the Commission issued a proposal for a regulation amending directive 2001/18 (European Commission 2010d). The regulation foresaw 'a legal base in the EU legal framework on GMOs to authorise Member States to restrict or prohibit the cultivation of GMOs that have been authorised at EU level in all or part of their territory' and applied to authorisations both under the horizontal safety directive 2001/18 as well as the food and feed regulation 1829/2003. The new Article 26b had become necessary, the Commission argued, 'to achieve the right balance between maintaining the EU system of authorisations based on scientific assessment of health and environmental risks and the need to grant freedom to Member States to address specific national, regional or local issues raised by the cultivation of GMOs' (European Commission 2010c: 7). In addressing the economic, social and environmental impact of the proposed law, the Commission finally expressed some difficulties with making predictions. Foreseeing any effects of the policy revision was problematic, the Commission caustically noted, as 'GMO cultivation in the EU has been very limited up to date' (European Commission 2010d).

Summary

When the European Commission first addressed biotechnology in the early 1980s, the political playing field was level. The Commission faced the choice between two different kinds of policies. It could emphasise harmonisation of sectoral legislation, foster an internal market for biotechnology products in the Community and create economies of scale sufficient to attract investment in biotechnology industries. Alternatively, the Commission could pursue a regulatory agenda that aimed at the highest levels of protection from risks specific to the technology and thus create a new field of EU regulatory policy. After it became evident that the two perceptions of the policy issue remained in conflict and that coordinated compromise was not within reach, the Commission adopted a more interventionist strategy. Emphasising technology-based risk prevention, the two central biotechnology safety directives of the 1990s not only imposed strict regulations. The shift in focus from market building to environmental protection also empowered a specific set of actors across the EU institutions, and caused specific procedures of decision making to apply. Sidelined by the adoption of this policy frame and its political repercussions, the political interest formation of the biotechnology industries interest groups and their advocates inside the Commission took on a new dynamic. In response to the legislative developments, European industries with stakes in the respective sectors formed a much more cohesive European lobby, the Senior Advisory Group on Biotechnology (SAGB). Their contacts with the policy makers responsible for industrial policy inside the Commission intensified. Because of their advocacy, the focus and emphasis of EU biotechnology policy, as adopted under the 1990 legislations, was called into question. When the agenda of the entire European Commission shifted towards the creation of the internal market only a few years later, they pushed the issue of biotechnology to the top of the EU agenda. The inclusion of biotechnology in the policy strategy of the Delors White Paper was a decisive step. The new coalition of biotechnology advocates managed to re-establish criteria of economic competitiveness as the official frame of reference for Commission biotechnology policy. Yet after the conflict over the definition of the issues became more virulent, the translation of the new frame into a reform of the EU legislative framework failed.

The legislative successes of the advocates of industrial and agricultural biotechnology were limited in scope and time. After a wrenching political process, the adoption of regulation on novel food temporarily managed to undermine the framework of the original safety directives. Especially the criterion of 'substantial equivalence' nearly eradicated the old regulatory approach in one of its central areas of application. But as the conflict over the framing of biotechnology seemed to turn to the biotechnology advocates' favour, the control of the Commission over the political dynamics in the field eroded. Most importantly, the persistent conflict over biotechnology policy facilitated the rise of the European Parliament. In 1995 the European Parliament used one of its first vetoes ever on the draft of the bio-technology patents directive. It became clear that the scope of the debate had extended to include new venues of political decision making and that the political ground had begun to shift. As a collective actor, the European Parliament initially rose in importance by sustaining the established conflict and contesting the new Commission approach. Yet eventually, a new strategy of framing the issue of biotechnology emerged. Set against the rising concern over food-related scandals throughout the mid-1990s, the advocates of regulatory safety shifted their arguments and the debate away from concerns of environmental protection to focus on consumer choice. The new policy frame funda-mentally transformed the conflict over the issues and facilitated a new legislative consensus – one of the world's strictest regulatory frame-works in the area of biotechnology. Controversy over the labelling of genetically modified products, a technical issue largely separate from the more fundamental questions of product authorisation, was the turning point. The new argument effectively twisted the political ratio-nale of biotechnology advocates around. Stringent regulation was no longer the opposite of a progressive, industry-friendly biotechnology policy. It became its precondition, an integral part of the strategy to create trust in a European market for biotechnology products. As a new dimension of the issue entered the debate with the focus on consumer protection, the established conflict over biotechnology policy collapsed. It ceased to unite the proponents of biotechnology deregulation, their voice faded out and institutional turf battles inside the Commission subsided. By the end of the decade, the fluidity of EU policy contestation had revealed its erosive impact. In 2001, the adoption of the revised, extended and substantially more stringent biotechnology safety directive, which replaced the original law from 1990, as well as the adoption of extensive follow-up legislation cemented the new policy frame. All ground gained in the interim period under the economic competitiveness frame was lost. 'The reframing [of EU biotechnology policy] in terms of the Delors White paper', one senior Commission official remarked in summary of political developments that spanned over 20 years, 'has totally failed' (Interview, European Commission, March 2005).

6
The Framing of EU Biotechnology

Over roughly two decades EU biotechnology policy turned from a regulatory nether land into one of the most volatile and bitterly contested policy conflicts in the European Union. How did this happen? Looking at the architecture of the policy conflict and the way in which the framing and reframing of the issues reconfigured the interests at stake sheds new light on the policy dynamics at play in the EU. Yet the analysis also raises some new questions. If policy framing was driving the developments in the late 1980s and early 1990s and allowed the Commission to outmanoeuvre political opposition, why did the reframing of biotechnology policy fail when the Commission tried to make the issues a part of its new economic policy agenda? How did policy initiatives that conformed to the original interests of member state governments and the biotechnology industries turn into almost the complete opposite of what the Commission set out to do? In addressing these questions, the analysis will focus on how the flow of the issues changed the structure of the conflict and affected policy dynamics that proved increasingly difficult to consolidate. After years of framing and counter-framing, the biotechnology conflict space was politically charged beyond the level of what the EU policy making system was designed to contain. In the end, the framing of the issues was transformed yet again and expanded the policy conflict to support a vast new alliance of interests.

In order to explain the initial framing success and the subsequent dynamics of reframing and counter-framing, the analysis must first take a broader look at the interplay of frame conflicts and the organisational structure of EU politics. This allows the analysis to address questions of how issue definitions affect the contours and allocation of policy jurisdiction, the choice of institutional channels and the applicable rules of decision making. Following Baumgartner and Jones (1991, 2002), the findings presented here support the view that the institutional channels of decision making, or policy venues, played a pivotal role and exerted a bias, both in terms of the dominant policy outlook and in terms of the set of actors that were empowered to take decisions (e.g. Baumgartner and Jones 1991: 1047). The analysis highlights how the way in which issues are defined during the policy formulation stage can affect institutional determinants of policy making by influencing the administrative and legislative allocation of responsibility. From this perspective, the organisational structure of government can be seen as a set of powerful intermediary factors acting upon policy outcomes or, to summarise the theoretical point, 'institutions are fundamentally endogenous to the policy process' (Baumgartner and Jones 2002: 4). At the same time, however, the empirical analysis also highlights how institutional responsibilities for separate issues of biotechnology remained fragmented and how the processing of ideas and demands continued to reflect the multifaceted nature of the issue. Attempts at framing and reframing the issues failed to reconcile conflicting perceptions of the issues at stake and failed effectively to silence or sideline political opposition. Institutionally entrenched policy subsystems with a lasting grip on the issues never materialised in the way many standard accounts of policy making would suggest. How the resulting issue politicisation affected the policy dynamics over time is addressed in more detail in the second part of this chapter.

The architecture of a policy conflict

The dimensionality of the debate

The first step in solving the puzzle of EU biotechnology policy is to explain its unusual potential for reversal. Over the first ten years, the Commission's biotechnology agenda shifted from environmental safety regulation to an economic policy approach with little concern for the rationale or substantive choices of the Commission's previous initiatives. Yet the Commission pursued its evolving policy agenda with some remarkable success. One way to analyse these twists and turns is by asking how the dimensionality of the policy debate changed in the process and how the framing and counter-framing of the issues caused different dimensions of the policy conflict to come to the fore.

As the discussion of the earliest Commission communications on biotechnology has shown, a twofold emphasis on the underlying dimensions of economic competitiveness and environmental safety prevailed from the early stages of policy making. Over many years, the advocacy of the two corresponding policy frames continued to highlight the mutually incompatible interests at stake and facilitated increasing tensions inside the Commission. As a brokered compromise was unlikely to emerge and policy coordination had proven ineffective, the Commission's political leadership chose one frame over the other. Initially, the Commission chose to push for biotechnology safety regulation. The resulting agenda showed little concern for the interests expressed by member states, scientists or the biotechnology industries, but it allowed the Commission to vastly expand its new competences under the environmental chapter of the Single European Act. DG Environment was assigned the co-chairmanship of the Biotechnology Regulations Inter-Service Committee (BRIC). As a result of the sub-committee's influence over the legislative drafting process, the new policy frame became institutionally anchored for the first time. The policy approach gained additional momentum with the Commission's subsequent decision to propose horizontal legislation to the Environment Council, which was the single decisive legislative body at that time. Unsurprisingly, environment ministers proved receptive to the Commission's chosen policy outlook. While the Environment Council was empowered in part as a result of the way in which the policy issues were framed, this legislative body now maintained the initial momentum and secured the adoption of a strict and encompassing regulatory approach. Shortly thereafter, however, challenges launched by disenfranchised proponents of the industrial biotechnologies became more organised and the advocacy of alternative problem perceptions inside the European Commission gained ground. As new dimensions of the policy issues became more salient, policy venues shifted in response. The combined effects facilitated an abrupt policy shift (see Baumgartner 2007: 484). This is what happened with the adaptation of the Delors White Paper and the move to revise or replace biotechnology safety regulations with sectoral legislation, drafted chiefly by DG Industry. During this period, political conflict over biotechnology policy was still defined by the same two issue dimensions, just as the advocacy for both frames of reference never ceased after the reframing of the issues. In contrast to the first phase of issue framing and policy making, however, the economic dimension now dominated, and a new set of actors and interests took centre stage. The fact that the adversarial confrontation over the framing of the issues continued at this stage of the policy debate also reflects the largely futile attempts by DG Research and the biotechnology industries to influence the structure of the policy conflict more fundamentally. More than any other actor, DG Research had made it its mission to counter the notion that the risks associated with biotechnology were as serious or uncertain as to merit stringent regulatory intervention and justify potentially stifling policy regimes across areas as broad and diverse as basic scientific research, pharmaceuticals and food production. While DG Research also took positions concerning the specific questions under consideration at various points in time, its central message was that both dimensions of the policy debate were not nearly as mutually exclusive as they were perceived or portrayed to be. Yet these argumentative interventions, and the evidence provided to back them, had little impact. The dimensionality of the debate remained stable over many years and when it finally changed, the biotechnology advocates played no part in steering the developments.

In this context, the conflict over food labelling illustrates better than any other single episode of EU biotechnology policy making how the flow and structure of the issues can recast policy dynamics. The biotechnology advocates had just won a major victory with the adoption of the novel food law when biotechnology was pushed back to the top of the Commission agenda. Open internal conflict over food labelling was clearly fuelled by jurisdictional disputes and legislative ambiguities. But more importantly, the question of labelling caused a shift in focus to consumer choice. In the process, the labelling dispute transformed the dimensionality of the debate and redrew the lines of conflict. The spotlight on the potential exposure to genetically modified food, feed and seed triggered retailer boycotts and consumer resistance. By 1999, the remaining political support for the economic biotechnology agenda in the Council was eroded and agricultural biotechnology in the EU practically came to a standstill. With the future of the entire sector hanging in the balance, biotechnology advocates realised that little could be gained from pressuring governments and the Commission to revert back to a policy agenda of deregulation and market harmonisation. The political chances to see further policy reform along the lines envisioned in the Delors White Paper appeared marginal. At the same time, the problem had changed from an economic perspective. Until now, the primary concern of biotechnology advocates had been to harmonise product legislation, lower the regulatory burden, simplify product authorisations, limit risk assessment both procedurally and substantively, and generally create legislative certainty so as to encourage investment in research and development. Although all of these concerns were still important in the mid and late 1990s and remained far from being resolved in the industries' favour, the most salient question now was marketing. Biotechnology industries needed to win over retailers and consumers in order to make the technology profitable. While the core policy objectives of biotechnology advocates had changed little, they had to concede the battle over the structure of debate. The focus on economic competitiveness was now increasingly used to justify the expansion of the same regulatory framework that advocates of industrial and agricultural biotechnology had originally set out to overcome. The prospects of EU biotechnology were portrayed as hinging on consumer trust and protection, which in turn required a revised and expanded regulatory framework, closely in line with regulatory safety provisions previously advocated under the environmental safety frame. As a result of the new dimensionality of the debate, the interests of the biotechnology advocates practically turned against themselves and the policy conflict collapsed. Correspondingly, as pointed out in the empirical discussion, towards the end of the 1990s the biotechnology lobby effectively rested its case. The flow of the issues and the transformation of the policy conflict had effectively disabled a once formidable alliance in supranational policy making while the new frame of consumer protection helped the advocates of stricter regulation to capture an ever expanding coalition of interests from across the EU institutions and beyond.

In sum, the first two phases of EU biotechnology policy were characterised mainly by a shifting policy emphasis within a stable conflict space. During this period, the dimensionality of the policy debate remained almost constant and opposing political alliances formed at the supranational level to compete over the right to define what was at stake in the issues. Repeated framing and reframing further fuelled the conflict. The emergence of some exceptionally well-coordinated campaigns by European environmental activists especially illustrates the resulting potential for political mobilisation. But the more consequential policy dynamic was not the ensuing battle of attrition. Instead, the creation of the consumer choice dimension brought about a displacement of the policy conflict – a flank attack by a bigger, collateral conflict that shifted the lines of political contestation (Schattschneider 1957: 939). After EU biotechnology policy had been driven and defined by the conflict over two conflicting issue dimensions, the final period in contrast was characterised primarily by the transformation of the dimensionality of the conflict itself.

Framing and counter-framing

The initial framing of biotechnology policy in the EU was a political success story from the viewpoint of the European Commission. The horizontal safety frame created a new policy field and helped to establish, legitimise and expand supranational policy competences where none had existed previously. Both in substantive and institutional terms, the notion of technology-specific regulation and the environmental policy focus was only one out of a number of alternative choices. EU regulation of safety standards for products was more established and widely considered to be sufficient and desirable. On this question, there was agreement among affected national interests, the scientific community and the majority of Commission services. The choice of a technology-based regulatory policy approach thus counts as a conscious political decision by the Commission. Its only advocate during the early days of EU biotechnology policy making, DG Environment, was internally isolated and institutionally weak as it remained without a treaty base throughout the initial phase of policy formulation. Nonetheless it was made *chef de file* and given almost free reign in the drafting process. While the adoption of the horizontal safety frame allowed the Commission to vastly expand its competences and establish its role in environmental policy making, this frame was also met with most resistance. But the Commission managed to disaggregate industrial interests and bypassed existing national reservations by channelling the issue through the Environment Council. Industry interest representation at the national level was predominantly organised according to economic sectors. Coherent positions on technology-based regulation were not readily available. Moreover, interest organisations were often split internally over the issue. As a result, an organised voice of the affected industries failed to form and the Council of Environmental Ministers was free from adverse pressure when it passed the two biotechnology regulations.

The decade of the 1990s hence began with the creation of a new supranational policy field. The reorganisation of industry interests happened in response to the political decisions at the European level, not only in the sense that industry representation took shape after only the two directives were adopted. In following the horizontal, technology-based approach of the new EU legislative framework, the biotechnology industries also adapted organisationally to the supranational policy design. Interest group restructuring graphically illustrates how early issue framing influenced the biotechnology policy space for years to come. Intriguingly, however, during the final stages of the legislative process that led to the adoption of highly restrictive regulations in 1990, the Commission was already engaged with the newly formed industry group and just one year later drew heavily upon their input in the drafting process for legislative reform proposals. Clearly, the Commission's interest in the field of biotechnology was not nearly as consistent in terms of substantive policy choices as it was in terms of its political interest in creating and shaping the policy field itself. From this perspective, the establishment of biotechnology interest representation at the EU level was a major step forward. Irrespective of the overt opposition the biotechnology industries expressed to the new policy regime, cooperation with the interest group was actively sought by the Commission. By successfully initiating an alliance of some of the largest chemical and pharmaceutical companies in Europe based on their use of a specific technology, the Commission also created the interest organisation and the policy demand it later exploited to legitimise the adoption of the economic competitiveness frame. The Senior Advisory Group with its limited membership and high level of internal cohesion was considered a case of particularly successful and potent adaptation of interest representation to the political system of the EU (e.g. Greenwood and Ronit 1994). Two years after the adoption of the biotechnology safety directives, Greenwood and Ronit (1992: 93) even anticipated that the industry group could transform itself into a Brussels-based private interest government with substantial self-regulatory competence. Instead, after a few short years and long before the dramatic revision of the regulatory framework, the Senior Advisory Group was converted back into a more traditional European interest association, which reached out to national industry groups as well as individual companies.

The main success of the biotechnology industries was in many ways its involvement in the case of the revised directive on contained use, which regulated the use of genetically modified microorganisms in laboratory research. Ironically, however, this success harmed the industry group politically and partially undermined its organisational cohesion. With the amendment of the directive on contained use in 1998 and, to a lesser extent, the eventual adoption of the biotechnology patent directive in the same year, the pharmaceutical industry secured two of its central legislative objectives in the field of biotechnology. The amendment of the contained use directive substantially relaxed safety regulations for laboratory research and streamlined administrative procedures. The biotechnology patents directive harmonised the legal protection of biotechnological inventions across the EU. As a result, one of the most potent sectoral interests in biotechnology was effectively free to withdraw and largely escaped the turbulent policy debates that followed. The early successes had in fact left the biotechnology industries with a weakened representation. After some of its sectoral interests had been served, the scope of the conflict over biotechnology had been reduced and, as a result, the direction of the conflict also changed.

The review of events thus indicates that the Commission was initially able to control the agenda, exploit the lines of conflict in the policy formulation process and facilitate policy adoption and revisions by framing and reframing the issues. But the repeated policy reframing also left the debate politically charged. Policy frames always remain contestable (Riker 1995: 34; Jones and Baumgartner 2002: 298). In the field of EU biotechnology, no single actor or coalition retained control over the dynamics for the entire period. The Commission first succeeded with the establishment of the environmental safety frame and the implementation of the original regulatory regime in 1990. The interim successes won in the wake of the Delors White Paper, especially the novel food regulation, however came under intense scrutiny as soon as the laws were adopted. After the novel food law passed, biotechnology advocates clearly failed to oppose counter-framing attempts and eventually lost control over the policy process. With the issues high on the agenda and political pressure rising both in the Parliament and the Council, a new supranational coalition with interests closely aligned to the original safety frame assumed control over biotechnology policy formulation. Instead of implementing the objectives advanced by the Commission in the early 1990s, the resulting policies substantially expanded the scope of the regulatory framework.

Reframing failure

The strategy to reframe the policy issues in the mid-1990s and reintroduce them on the political agenda as a cornerstone of the new agenda for economic competitiveness and growth largely failed. The Commission secured some successes with the revision of the contained use directive and the eventual adoption of the biotechnology patents directive, following Parliament's initial veto in 1995. But the cornerstone of the new approach, the novel food regulation, was repealed only a few years after it came into force, and some of its central provisions were under attack as soon as the policy was adopted. Why did the reframing strategy fail to wield lasting influence over the policy process? Especially given the fact that the reframing of the issues took place in close cooperation with the affected industries and constituted a major agenda setting initiative by the Commission during the Delors presidency, the ineffectiveness of the new emphasis on economic competitiveness in restructuring the supranational policy dynamics is surprising. Policy reframing was furthermore sustained by the fact that issues of food production, which dominated much of this period, were already in part under the direction of DG Industry. The roster of participants in the Council and, to a lesser extent, Parliament, was therefore different from when the 1990 legislation was agreed. All of these factors should have helped to create and sustain policy dynamics in line with the reframing of the issues. They did not.

As outlined above, the novel food regulation substantially changed the character of the original regulatory framework. The law set out significantly simplified authorisation procedures specifically for food products and avoided the stigmatisation of genetically modified food by adopting lax labelling schemes. The notion of horizontal safety legislation was further diluted by the introduction of the criterion of 'substantial equivalence', which was non-exclusive in terms of organic and genetically manipulated food products and functioned, in the view of its critics, as a backdoor for genetically modified food to enter the European market. During the discussion of the regulation in the Parliament, the effect of the shift of attention away from safety standards and the greater emphasis on the economic prospects of biotechnology initially dominated the debate. The Environment Committee still held fast to its own frame and persisted in proposing far-reaching changes to the law, most importantly on the issue of mandatory labelling. This appears to confirm the assumption that institutionally embedded frames retain high levels of stability even under increasing pressure to integrate new evaluative dimensions. Yet the institutional effects were limited to the Environment Committee and did not affect the vote of the assembly to the same extent. The floor votes repeatedly dismissed the most substantive changes proposed by the committee and eventually accepted less safety-oriented provisions.

Nevertheless, the victory of the advocates of a new approach to EU biotechnology over the environmental alliance did not last. The adoption of a new policy frame supported by DG Industry did not reshape the existing lines of policy conflict. Instead, it revived the existing conflict space. The drive for deregulation kept the concern for biotechnology safety on the EU political agenda and, over time, this enabled other supranational institutions, notably Parliament, to reconsider its position. Unwittingly, the Commission's pursuit of a new agenda helped rather than hindered in this process. When DG Industry began drafting the novel food policy during the mid 1990s, it dominated most of the Commission's activity. The interests that had previously allied with the Commission in pursuit of environmental safety and health protection lost their primary policy venue. As a result, environmental interest groups were free to exploit the emerging conflict and take it to a European Parliament eager to exercise its newly gained powers under the co-decision procedure. Just as the Commission had successfully deployed the horizontal policy approach to circumvent adverse sectoral interests and empower the newly emerging biotechnology associations at the EU level, the adoption of the economic competitiveness frame provided the rationale behind the increasing engagement of the Parliament with environmentalists. Environmental interest groups had successfully established relations with Parliament at the time of the 1995 veto of the biotechnology patents directive. The Commission's plans for the upcoming revision of the deliberate release directive now enabled the interest groups to redeploy their resources, expand the scope of the debate and turn the fight for environmental safety and consumer protection, as one industry lobbyist observed (Interview, Brussels, March 2005), into the 'raison d'être' of the European Parliament. The upcoming review and reform of the original horizontal safety policy sealed the fate of the Commission's new biotechnology agenda. Directive 90/220 on the deliberate release and marketing of genetically modified organisms came under review in 1996, three years after the publication of the Delors White Paper. Its massive regulatory scope in comparison to the much more specific directive on contained use or the novel food regulation made the deliberate release directive the most central law in the policy field. Moreover, all sectoral legislation, such as the novel food regulation, was linked to the directive in terms of environmental risk assessment. While the exact interpretation of this linkage changed slightly over time, it became increasingly relevant and was eventually written into the text of the revised directive. Minimal standards for risk assessment of genetically modified organisms were to be defined by the horizontal directive and thus retroactively affected the rest of the EU regulatory framework. The centrality of the policy only increased as the Environment Commissioner began issuing labelling rules under the horizontal directive at a time when the labelling issue was hotly disputed at the European level and a more general labelling regime was not yet in place. In short, the safety directive was the key to turning EU biotechnology policy around and implementing changes across the entire regulatory framework. Yet instead of revising the directive in accordance with the reviews and reports written between 1996 and 1997, the Commission lost control over the policy dynamic. Eventually, the Council and Parliament pressured the Commission to adopt a revised and extended regulatory framework that even exceeded the policy objectives of the Environment Commissioner and transformed EU biotechnology legislation into not only almost perfectly harmonised EU law, but also into the most stringent safety regulation in the area of biotechnology worldwide. The novel food regulation, the key success won under the economic competitiveness frame, could not be shielded from these policy revisions, and all of its provisions pertaining to genetically modified food were repealed by the more stringent policies adopted in 2003.

Framed by public opinion?

In explaining the Commission's failure to reframe biotechnology, public opinion has been widely ascribed the role of the 'joker in the pack' (Cantley 1995: 656). According to this argument, had it not been for the outbreak of BSE (bovine spongiform encephalopathy) and other food-related scandals over the course of the 1990s, the strategy of the Commission to reframe biotechnology in terms of its positive effects on Europe's economic competitiveness might have prevailed. While the link between traditional food scandals and the regulation of EU agricultural biotechnology is less than straightforward, these external events may nonetheless have caused political dynamics that ran counter to the Commission's intentions. It appears easy to substantiate this analysis. Public support for genetically modified crops and food was dwindling throughout the period of deliberation over the reform of EU biotechnology policy. In the years from 1996 to 1999, when the BSE crisis shook Europe's food sector, support for genetically modified food across Europe dropped from 45 to 36 per cent, while outright opposition rose from 39 to 52 per cent (European Commission 2003b: 16). Contrary to this line of reasoning, however, a look at the chronology of events indicates that the BSE crisis did not in fact have a decisive effect on the eventual failure to reframe EU biotechnology policy. The advisory committee of the Creutzfeldt-Jakob Disease surveillance unit at Western General Hospital in Edinburgh revealed that BSE had apparently jumped to humans on 20 March 1996. This was 10 months before both Parliament and Council adopted the novel food regulation under the co-decision procedure. The very law that was the centrepiece of the new economic agenda in the area of agricultural biotechnology, and that was heavily criticised at the time for allowing biotechnological food products to bypass the safety standards set up by the original regulatory framework, was hence adopted by both chambers of the EU legislature just around the time when the food scare in the EU peaked and pictures of smouldering cattle could be seen on television screens across the Union. It therefore appears unlikely that the food scare of the 1990s was the decisive external factor that powerfully overrode all other political dynamics and cancelled out the Commission's agenda setting power at a critical time. Short of providing strong evidence of a direct effect on the eventual failure of the European Commission to reframe biotechnology policy, changes in public opinion can instead be integrated from a framing perspective and their effects endogenised. This interpretation would place emphasis on the translation of external factors into political dynamics already at work at the supranational level and ask how they were transformed, rather than superseded.

Sidelined by institutional reform?

Just as external events could have intervened decisively in the course of events, the changes in the institutional decision making rules in the EU over the period included in the empirical analysis might have affected the Commission's ability to successfully frame policy initiatives. Over the course of the first two decades of EU biotechnology policy making, the formal role of the Parliament grew with every treaty revision. With the adoption of the Single European Act in 1986, the consultation procedure, under which the Council was the principal legislative body, was supplemented by the cooperation procedure and Parliament was given more far-reaching formal powers. The co-decision procedure, introduced by the Maastricht Treaty in 1992, and simplified and extended by the Amsterdam Treaty in 1997, gave the Parliament and the Council equal legislative rights (see Nugent 1998 for a concise overview). The view that the new co-decision procedure diminished the Commission's influence over EU policy making is widespread (e.g. Tsebelis and Garrett 2000), and the legislative Commission's demise could easily be counted as a sufficiently strong effect to blur the analysis of independent framing effects in a cross-case comparison of law-making before and after the procedure's introduction. Furthermore, the claim that the rules and reforms of inter-institutional decision making in the EU make all the difference has particular plausibility in the case of biotechnology legislation, the policy field in which the European Parliament used its veto power to block the adoption of biotechnology patents in 1995 (Rittberger 2000). But were the framing successes of the Commission in fact limited to those legislative initiatives that could be pursued without substantial involvement of the European Parliament? In other words, did the increase of policy oversight and political competition caused by these fundamental institutional reforms transform the European political playing field in such a way as to diminish or even eliminate the scope for policy framing as a viable policy making strategy? It seems they did not. A straightforward institutional explanation of the reform of EU biotechnology policy falls short of meeting even the most obvious tests. Again, the novel food regulation serves as a case in point. The law was adopted in 1997 under the co-decision procedure. The Parliament had long since used its veto power and established itself as an equal legislature together with the Council. When the Commission presented the novel food regulation, the members of the European Parliament were therefore presented with the chance to sink the Commission's new biotechnology agenda right from the start. Yet instead, the novel food regulation passed the European Parliament largely intact and the biotechnology advocates secured a substantial (albeit short-lived) legislative victory. As noted above, the empirical case in fact reveals strong effects of policy reframing on the votes in the Parliament. The responsible environmental committee, in outright opposition to the new law, continued to advocate the original safety frame and attempted to block the passage of the novel food regulation in the form in which it was later adopted. But the responsible parliamentary committee for this policy area was outvoted in the plenary. This is even more surprising since party politics, in addition to the committee's recommendation, would have indicated early opposition to the Commission initiative. The Social Democrats still formed the strongest group in Parliament at the time, provided the rapporteur who handled the legislative file in the case, and traditionally allied with the Green group on biotechnology issues. Yet not even the intense lobbying by the environmental interest groups that had successfully established contacts with the Parliament during the biotechnology patents campaign in the same period, nor the last-minute attempts by the members of the Green group to assemble enough votes for a renewed veto effort, managed to offset the policy dynamics. The plenary voted in line with the Commission's point of view that deregulation and economic interests had to take precedence over environmental safety concerns. Until the end, the environmental committee tried to stem the tide, but the Commission's reframing strategy had successfully eroded the committee's control over the issue and its backing in the assembly. The reframing strategy of the Commission proved robust enough to survive the reformed set-up of EU legislative politics and to undermine the responsible committee's sway over the plenary vote. Co-decision, in short, was not the problem.

Conflict and consolidation in EU policy making

The analysis has shown that the architecture of the policy conflict, the framing of the issues and the processing of the issues by different institutions were central to understanding the policy process. Yet the ways in which the reframing of the issues failed and instead resulted in a transformation of the debate that disempowered the coalition of biotechnology advocates remains puzzling. If the framing of the issues of EU biotechnology was such a decisive political move in the late 1980s and allowed the Commission to bypass almost unilateral opposition to its policy initiatives, why did policy reframing fail in the 1990s, when the Commission tried to make biotechnology a cornerstone of Europe's new economic policy agenda? What accounts for the fact that policy initiatives that conformed to the original interests of member states and the affected industries alike turned into almost the complete opposite of what the Commission set out to do? Focusing on the interlocking effects of policy framing and issue politicisation in the EU, the following analysis seeks to highlight the fact that the political system of the EU failed to provide mechanisms of political consolidation that were strong enough to shield framing effects from persistent contestation. While policy framing and reframing has been shown to structure policy conflicts, this theoretical perspective also perceives of consolidation as the flip side of political conflict (Schattschneider 1960: 60). The empirical analysis indicates, however, that increasing politicisation of policy conflicts in the EU can instead lead to recurring policy shifts as the issues become more difficult to contain and the policy conflict continues to transform. In the case of the EU biotechnology policy, the conflict space only stabilised after the scope of the conflict had expanded dramatically over a period of more than twenty years. The following discussion takes a closer look at these political processes and proposes some tentative findings concerning the nature and function of issue politicisation in the political system of the EU.

Issues, interests and compartmentalisation

The initial adoption of the safety frame by the European Commission, both in terms of its substantive focus and in terms of the assignment of institutional responsibilities inside the Commission, was not a reaction to external political pressure. Political pressure on the Commission was generally weak in the early days of EU biotechnology policy making. Where it existed, often as input explicitly solicited by the Commission, everything indicated that a stringent regulatory approach would run counter to existing political demand as expressed by member state governments and by the scientists and industries affected by the policy choices. Yet despite its appearance, the choice of the horizontal safety frame was not so much a decision against the economic exploitation of European biotechnology as it was a choice for a new supranational policy field. To adapt product legislation and to emphasise market-building and harmonisation would not have allowed the Commission to bypass some of the strongest domestic industries and their existing interest organisation at the national level. In other words, biotechnology product legislation would have been made in the Council. A distinctly supranational policy field that went beyond the aggregation of national preferences would not have emerged. That the horizontal safety frame did not in fact stem from a pervasive anti-biotechnology attitude inside the Commission became obvious as events unfolded. Immediately after the adoption of the two 1990s directives, the Commission turned around and established intensive cooperation with the newly founded biotechnology industries groups in Brussels. The Commission's political leadership was clearly backing the new agenda, as the adoption of the Delors White Paper showed. This White Paper virtually copied its ideas from earlier DG Industry communications, which in turn drew heavily on language provided by the newly founded biotechnology advocacy group. Yet the abrupt policy shifts that were necessary to create and exploit interest formation at the European level also caused friction among the Commission Directorates-General. Different Commission services not only processed issues from their respective policy area independently from one another (see Cram 1994). The Commission also processed individual issues from the same field of policy making simultaneously in different DGs and often two or more competing DGs processed the very same issue at the same time. The extent to which the complex compartmentalisation and processing of the issues in the Commission remained independent from the framing and reframing of the issues was often stunning.

While a multiplicity of organisational access points was vital for the Commission's ability to reframe the biotechnology debate, the prevailing tendency for them to decompose into competing organisational arenas as soon as issues arose on the EU agenda also counteracted the Commission's pursuit of a single coherent line of action. Coordination across the Commission services failed repeatedly, illustrating the difficulties in overcoming framing differences through bargaining, argumentation or evidentiary policy discourse. After the Biotechnology Steering Committee was abolished, the Biotechnology Coordination Committee took over in 1989. The frequency of its meetings declined almost immediately after its foundation and ended in the renewed abolition of the coordination committee and the reinstatement of the Biotechnology Steering Committee in 2003. During this period, the Commission's biotechnology committees typically met after major legislation had been adopted, rather than before, to discuss the newly emerging pressures and political demands that resulted from recently adopted policy at the EU level.

Given the high level of internal contestation, the college of Commissioners twice faced significant policy choices that had proven impossible to resolve and demanded up or down votes. The first case was the proposal for the revised directive on contained use. The responsible Commissioner proposed tightening regulatory rules after an attempt at coordinating policy across the different services had failed. But her colleagues voted her legislative initiative down and instead adopted an alternative proposal, tabled by the Industry Commissioner. In the second case, the President of the Commission was outvoted in his attempt to establish a new labelling regime for genetically modified food. Though he had personally assumed policy responsibility for the draft, the college deflected his intervention and adopted a counter-proposal drafted by DG Agriculture in cooperation with DGs Environment and Consumer Protection. As outlined earlier, parallel processing of contentious and multi-faceted policy issues typically continues until conflicts push the issue higher up the decision making hierarchy. As the highest decision making unit in the European Commission, the College of Commissioners correspondingly had the 'coordination function that is the primary responsibility of these levels' (Simon 1973: 271). Yet in the absence of any additional, salient organising principle unique to this level of decision making in the Commission, such as political ideology or territorial representation, the Commissioners' preferences over the conflicting frames in biotechnology policy remained volatile.

Speaking to this aspect of the empirical puzzle, as outlined in more detail above, research on political behaviour of Commissioners has shown more generally that this body's highest political representatives shift back and forth between different representational logics, sometimes invoking the Commission's institutional interests as a collective actor, sometimes following political rationalities based on their respective policy portfolio, country of origin, or party membership. This research indicates that portfolio organisation inside the Commission is often vital (e.g. Trondal 2007b: 963), as it was at several junctures during the process of biotechnology making, but none of the representational principles which find expression in the Commission's internal organisation served as reliable institutional determinants of political behaviour over time. In sum, just as the turn of events all but eradicated any hope of making sense of the European Commission as a unitary actor, it appeared true for most EU political actors that 'representation involves balancing multiple competing roles in different situations at different times' (Trondal 2008: 447).

While political organisation clearly matters, the larger point here is that in the case of the EU it is often not sufficient to 'unpack the basic organisational characteristics of the institutions within which individuals interact' (Egeberg 2004: 201) in order to unearth the decisive mechanisms of policy formation and choice. Once confronted with complex and salient policy issues, a political system riddled with conflicting organisational logics and incoherently reconciled representational functions will often fail to produce political mechanisms strong enough to reach or sustain policy equilibria even at the intra-institutional levels. 'The very fact that there is no perspective for the resolution of these conflicts', Christiansen (1997: 76) contends, 'is what makes the European Union so distinctive'. In short, the Commission proved exceptionally effective in creating a multidimensional, supranational conflict space in the area of biotechnology policy. It was less effective in containing or ordering the conflicting dimensions throughout the policy process. As biotechnology policy fluctuated back and forth between periods in which different issue dimensions were salient and different actors inside the Commission were in charge of the issues, the inter-institutional dynamics of EU politics became increasingly relevant.

Supranational political competition

During the 1990s, after the Commission had switched to defining biotechnology as economic and industrial policy, the Parliament had no intrinsic interest in pursuing strict environmental legislation. At the time, the rationale behind the Commission's attempt to stimulate economic growth, increase European competitiveness and provide for more jobs was evident and convincing given the economic development in most EU member countries. The notion of a highly visible EU economic policy package certainly did not interfere with the Parliament's ambitious goals for supranational policy making in the EU. Furthermore, the number of MEPs who were actively involved in shaping Parliament's majority positions on biotechnology never exceeded a dozen (Interview, Brussels, March 2005). But despite all this, the European Parliament made it its mission to prevent the economic reform agenda from affecting the field of biotechnology – one out of three policy fields singled out by the Commission White Paper for immediate action. Why was the biotechnology safety frame so rarely contested among the members of Parliament and eventually dominated this political body's policy positions so entirely?

Even during the height of the food scare in Europe, EU parliamentarians were aware that links between traditional food epidemics such as BSE and safety issues concerning genetically modified food and feed were more political than substantive. Yet parliamentarians eagerly adopted the position that existing public perception called for increasingly stringent regulation in the field of biotechnology. In terms of substantive policy, the Parliament thus allied itself with environmental interests early on and later pioneered the move to highlight consumer protection. When the Commission began to reframe biotechnology as economic policy, the arising conflict offered not only a perfect opportunity for Parliament to exploit the issues and to form an alliance with increasingly sidelined environmental interests. It also legitimised institutional stand-offs between the Parliament and the Commission, such as over biotechnology patents in 1995, from which the chamber emerged as a strengthened legislative body. Lacking a clearly identifiable constituency that would pressure the members of Parliament to advance policy objectives based on their membership of one of the political groups or given their roles as national representatives, the Parliament as a whole was free to pursue its collective institutional self-interest. The alliance with environmentalists and consumer protection activists fulfilled this purpose perfectly. The failure of industry associations to identify the Parliament as a political partner until after 1995 certainly meant that the Parliament's choices were limited. More evidently, however, a different alliance would not have facilitated the exploitation of the policy conflict in the same way. During this time, the passage of the novel food regulation in 1997 exemplifies that the Parliament was initially receptive to reframing dynamics and willing to support legislation that compromised the original horizontal framework of biotechnology safety regulation. With few exceptions, however, the reframing of the issues did not spill over into competition among the parliamentary committees. Nor did the reframing entice ideological partisanship to come to the fore. Since the policy shift did not translate into mechanisms that could sustain the momentum inside the Parliament, interest formation remained volatile and multiple representational roles remained in play. As soon as the economic reframing became contested, the Parliament quickly turned into the decisive institutional venue for actors pushing for encompassing reform and even stricter regulatory standards.

The extent to which dynamics of supranational competition between the key institutional players trumped alternative mechanisms of interest formation on the intra-institutional level during this period is surprising. Parliaments with weak party discipline and far-reaching committee responsibilities such as is the case in the EU are generally considered more likely to delegate policy responsibilities to their committees and sub-committees. In the case of complex and multidimensional policy issues, the expected outcome would thus be internal competition over the portfolio. Established examples of turf wars over legislative competencies (e.g. King 1994) highlight attempts by committees to impose their respective issue perceptions and expand their purview. In such cases, interest groups typically lobby competing committees to engage in these conflicts when other institutional venues prohibit them from gaining political leverage, much in line with the general thesis of Baumgartner and Jones (1991). Collective institutional frames of reference, in contrast, typically gain little impact on the way in which legislative bodies define their interests. In fact, in his work on heresthetics, Riker (1986: 106–113) cites as an example of exceptionally skilful and creative political framing the case of a US senator who wins a vote by invoking the constitutional role of the Senate in foreign policy making, thus reframing the issue under discussion in terms of institutional politics instead of substantive policy. In the case of the European Parliament, however, a collective institutional definition of interests prevailed almost systematically whenever the issues at stake where politically salient. This was true above all in the case of the veto of the biotechnology patents directive, which had been cleared in conciliation and feverishly backed by the rapporteur before the floor vote. But the responsible parliamentary committee saw its policy preferences outvoted in the plenary. In the other cases of biotechnology policy discussed earlier, rapporteur allocation in the Parliament further indicates that priority was given to those political groups who were in strongest opposition to the Commission's reframed biotechnology agenda, in most cases the Green group and the Socialists. The presumably more industry-leaning EPP, on the other hand, claimed the rapporteur function only twice: in the cases of the less salient revised directive on contained use of genetically modified microorganisms and in one of the follow-up legislations of 2003. In both cases, the policy conflicts had largely been settled by the time the first Commission proposal was issued. Furthermore, in the case of the revised directive on contained use, Parliament only had reduced institutional clout under the cooperation procedure, making the rapporteur assignment significantly less attractive. Even if conservative members of the European Parliament were more open to the economic competitiveness frame, it is clear that they did not choose to use the issues of biotechnology in order to sharpen their party-political profile. Open conflict with the dominant Green minority in the assembly would have allowed the conservative members of parliament to build networks with biotechnology interests at the European level. Yet few conservative members of Parliament ever engaged in EU biotechnology policy at all. From the start, the Green group and some Socialist deputies maintained a firm grip on the relevant committee assignments and they were never seriously challenged. This remained true even when, following the 1999 European elections, the conservative EPP represented the strongest political group in the Parliament. Remarkably, the repeal of the most central biotechnology directive, the law on the deliberate release and marketing of genetically modified organisms in 2001, took place in the wake of a conservative surge in the Parliament. But the effects of the election on the Parliament's positions on biotechnology were marginal at best. Turf battles between parliamentary committees remained just as rare. The discussion of the review of directive 90/220 on the deliberate release instead reveals that the parliamentary committee that was most likely to resist calls for more stringent regulation was at least partially unaware of the biotechnology industries' positions on the issues and apparently confused about the EU legislative framework in force at the time. Instead of challenging the prevailing drive for stricter standards, the Committee on Economic and Monetary Affairs and Industrial Policy innocently went along with the rest of the committee reports, stressing the concerns of the European public and proposing legislative measures to increase regulatory tightness. The biotechnology patents law is another stunning case in point. After the Parliament's veto of legislation on the protection of biotechnological inventions in 1995, the reintroduced biotechnology patents directive was one of the most closely watched pieces of legislation in the EU at the time. The focus on the legislative role of the Parliament had rarely been stronger. But the political competition among different committees and the competition over crucial committee assignments among the political parties were again limited. Instead of encountering competition from the other political factions, the Green group easily secured three out of six rapporteur positions in the various committees involved in the process. Moreover, the six committees appeared to have so little independent issue perception to bring to bear on their policy positions that members of the Green group in many cases successfully introduced identical amendments in all committees simultaneously.

Both the high degree of attention Parliament devoted to biotechnology legislation and the way biotechnology legislation was processed by the chamber appeared to be driven by factors other than internal contestation and was largely decoupled from intra-organisational dynamics. Instead, it appeared to reflect the institution's overall strategy of policy politicisation. The Parliament stirred up conflict over the issues because policy conflict served its collective institutional interest. This argument also holds if this logic is reversed. The case of the directive amending the regulation of the contained use of genetically modified microorganisms constitutes the least controversial case of policy revision included in this empirical analysis. The original law was coupled with the infinitely more contentious deliberate release directive, and both laws were adopted together in 1990. The contained use directive was therefore part of the original regulatory safety framework. As such, it had caused the affected pharmaceutical and chemicals industries a considerable headache, as research activities were hampered by the strict safety precautions and the extensive authorisation procedures that applied to the uses of modified microorganisms. Fully in line with its new economic agenda, the Commission pushed for deregulation and reform. Despite objectives that were clearly at odds with the regulatory philosophy advocated by the Parliament both before and after the reform of this law, the assembly expressed little interest in it. Processed under the cooperation procedure, Parliament's influence was more limited from the start. But the lack of willingness to engage either the Commission or the Council in a political fight, if only as a signal to the environmental lobby that Parliament stood firm while the Commission's new biotechnology agenda took shape, remained startling. The Green group and the Socialists did not object to a conservative member of the assembly, a trained scientist, taking over the rapporteur function in the environmental committee. Amendments proposed by the Parliament were so uncontroversial and constructive that they were mostly incorporated into the final text of the law. Had the aim of the environmental safety advocates in the Parliament been to inhibit European biotechnology industries from expanding their activities in research and development, the proposal for the revision of the contained use directive would have been a critical item on their agenda. But a piece of legislation that offered so few possibilities to manipulate the scope of the policy conflict or affect the dimensionality of the debate was, like the title of the legislation itself, of 'contained use' to the assembly. Biotechnology industries across the EU rejoiced, and the Parliament turned its focus on the upcoming review of the deliberate release directive – with very considerable energy and devastating consequences for the coalition of biotechnology advocates.

Policy volatility and politicisation

The preceding sections have highlighted the competing effects on the political behaviour of EU actors rooted in the inter-institutional and intra-institutional organisation of the EU political system. The analysis has shown that the EU supranational level provides for weak structural features to ensure that policy perception and interest access systematically shift across the institutions of the EU according to a common logic. Instead, the fragmentation of the EU political system and the multiple coexisting political rationalities of key supranational actors ensure that great inconsistency in the way policy is processed is the rule rather the exception, both within and across the institutions of the EU. Instead of adhering to behavioural patterns that would compensate for these structural weaknesses, as common references to the EU's consensual and technocratic policy making style would seem to imply, EU actors at times actively exploit and politicise the complexity of the EU policy process. Looking at the EU from a framing perspective enables the analysis to focus on the role of policy issues in these processes and reveals why the flow of the issues and the manipulation of the structure and salience of the underlying evaluative dimensions of the policy choices can turn into causal drivers of supranational policy making. This perspective also reflects well established findings in studies of both the Commission and the Parliament. Students of these institutions have long argued that the bodies depend both for information and legitimisation on the mobilisation of political demands at the supranational level. The European Commission in particular seeks, fosters and funds the influx of policy demands and uses its command of policy information to claim its own mandate or constituency, one that does not merely trickle up from the national levels. But while the 'Commission is open, far more open than most national administrations', Marks and McAdam (1999: 105) point out, 'it is highly selective'. The legitimisation and information rationale drives the Commission to ramp up policy discourses and frame issues in ways that legitimise policy interventions beyond the status quo. Analogously, the 'accessibility of the European Parliament for unconventional activity' (Marks and McAdam 1999: 106) is of course due in part to the chamber's weak cohesion in terms of party organisation, but it has more direct roots in the inter-institutional competition at the supranational level.

In the EU, Majone (2002: 327) notes, 'the prime theme of the internal political process was, and largely still is, the contest of autonomous institutions over the extent and security of their respective jurisdictional prerogatives'. Looking more explicitly at legislative

decision making, Peterson (2001: 300) equally stresses that EU political conflicts are traditionally inter-institutional rather than intergovernmental. Not unsurprisingly, students of the Parliament have
long argued that EU deputies often find it attractive to offer an arena
for those issues, interests and demands that are disenfranchised or
sidelined by a specific policy approach taken by the Commission.
Rather than mirroring the political bias of a Commission proposal,
the Parliament frequently thrives on challenging the prevailing policy
perceptions. Peters (1992: 92) was an early observer of the Parliament's
propensity to switch back and forth between a policy game and a
game of institutional politics. While the first game reflects how the
assembly's various fractions form a majoritarian position on a substantive question of policy choice, the second is concerned with the
Parliament's pursuit and exploitation of its ever increasing institutional prerogatives as a collective actor *vis-à-vis* the Commission
and the Council. Depending on which game prevails at which point
in time, legislative majorities in Parliament can easily change as its
members re-evaluate conflicting objectives and roles. The internal fragmentation, overlapping competencies and multiple logics of political
representation of supranational actors regularly ensure that alternative
issue definitions remain in play simultaneously. To gain competitive
advantages, both collectively and internally, supranational actors therefore have vested interests in drawing increasing numbers of incompatible issue perceptions and conflicting demands into the EU decision
making process. At the same time, this dynamic often conflicts with
the overall policy making logic of a political structure that ultimately
depends on stable inter-institutional agreement to see politically
sustainable policy change enacted.

Looking at politics as a dual dynamic of conflict and consolidation,
Schattschneider (1935: 288) contends that within limits, 'every regime
can choose and formulate the pressures to which it will be subjected'.
While the political motivations and the institutional possibilities to
seek out, create and sustain conflicting pressures in EU policy making
seem particularly abundant, the mechanisms of consolidation are comparatively weak and erratic. Given the multiple constituencies and
complex system of representation, whose perceptions and demands
should be reflected always remains contentious. At the same time,
'populist policy equilibria' (Jones 1994b: 158, 174–175) that reflect
mass preferences over policy issues are notoriously difficult to seek out

at the EU level. As a corollary, more selective forms of issue politicisation provide rich opportunities to exploit these structural weaknesses. Policy framing is at the heart of this political process because it captures decision making initiatives that seek to construct or reconfigure the way given political interests play out over a certain policy issue. While this process may unwittingly provide EU institutions with more policy feedback and enhance the scope of EU political representation, it also creates stresses that might exceed what the EU policy making system is designed to process (see Bartolini 2006: 56).

7
Conclusion

The central theoretical puzzle of this book concerned the ways in which policy framing affects the processing of political interests and ideas and their expression in policy choices. Giving policy issues short shrift in theoretical accounts of policy making has biased standard frameworks of analysis towards conceptualising the effects of institutional structures and the political organisation of interests as largely independent from the issues at stake. The question of how the flow and structure of policy issues influence policy dynamics was largely ignored. Looking at theories of EU policy making, this book opened by arguing why it makes sense to shift the focus on policy issues.

The EU political system is characterised by competing constituencies and contested competencies. Its policy process is riddled with conflicting organisational logics and incoherently reconciled representational functions. The definition of what is at stake in an issue often remains contentious. The resulting interplay of political forces is therefore largely endogenous to the process of supranational decision making. EU policy choices are difficult to understand unless the analysis takes full account of the procedural dynamics of policy formulation. How issues are packaged not only affects the scope and contours of institutional turf battles. It also impinges on the ability of political interests to mobilise effective representation at the various levels of the political system, because institutional prerogatives and political logics of EU policy making can easily change over time and across issues even within the same area of policy making. Instead of facing a stable, institutional opportunity structure or a single, easily co-opted bureaucratic monolith, organised interests in the EU often struggle with shifting points of access. Even established policy communities can easily lose their grip on a policy area as decisions pass through the policy making system and multiple actors compete to define the structure of the issues at stake at the various intra- and inter-institutional levels. Models that are premised on pre-existing dimensions of political competition in the EU therefore risk overemphasising the inter-institutional complexities of the EU while losing sight of the highly variable types of contestation within and between them. Similarly, preconceived notions of institutional agency tend to preclude the stunning tendencies of EU collective actors to decompose into competing organisational arenas from the analysis. Rather than seeking out new ways of accounting for such mechanisms *ex ante*, EU policy research needs to take account of the political construction of interests at the supranational level. By identifying the dimensions of policy debates as 'moving parts of the spatial model of politics' (Riker 1990: 46), the framing perspective can provide a much needed alternative to the problematic analysis of EU policy making as characterised by a stable, exogenous and low-dimensional political conflict space.

Policy framing is critical in this context because the framing of the issues structures the conflicts to which issues give rise. Policy makers not only puzzle over the right answers to difficult questions. The framing of the issues in the policy process also affects how policy issues allocate and reallocate power. This theoretical argument primarily builds on the work by Schattschneider (1957, 1960) and Simon (e.g. 1985, 1987). Intriguingly, however, the predominant conceptual and theoretical foundations of existing research on EU policy framing typically deflect attention from the analysis of the political construction of the European conflict space more than they can contribute to it. The most commonly referenced literature on policy framing used in EU policy analysis today investigates how policy controversies that are rooted in conflicting problem definitions can be overcome. This line of research is influenced heavily by the work of Rein and Schön (1991, 1993, 1996 and 1997; see Schön and Rein 1994 for the standard reference). Unlike other types of policy conflicts that can in principle be settled by recourse to facts and established rules, Rein and Schön (1991: 265, 267) define policy controversies as conflicts that can be solved when participants 'reflect on the frame-conflicts implicit in

their controversies and explore the potentials for their resolution'.
Based on an expansive conception of framing, Rein and Schön formulate a largely normative theory of policy deliberation that focuses
on the social processes through which frame-reflective policy discourse can help adversarial political actors resolve substantive conflicts of interest consensually. The analysis offer here concludes that
the scope for 'frame reflection' as envisioned by Schön and Rein
(1994) may be more limited than the routine referencing of their
work in studies of EU policy framing implies. In fact, little indicates
that trust-based communicative discourses prevail even inside the
Commission once policy issues rise on the EU agenda. More typically, opposing coalitions are found to shore up support and thereby
politicise policy controversies on the supranational level. As Scharpf
(2000: 777) warns with a view of the EU, in multi-actor systems
with weak accountability, 'policy outcomes are more affected by
incentives favoring cooperation or conflict'.

In this context, the preceding analysis suggests that one key
to understanding the policy dynamics at the EU level may lie in
exploring the interlocking effects of policy framing and EU politicisation. As the empirical discussion has shown, even technical policy
issues can become the linchpins of inter-institutional contestation
and unwittingly help to shift the centre of EU political gravity further
towards the supranational level. Since no other actor in EU politics
appears as uniquely positioned to exploit the resulting complexities
of EU policy making in pursuit of its own agenda, the analysis has
emphasised the Commission's role and focused on its ability to
sway other actors in the EU policy process. The results were mixed.
While some findings illustrate that the Commission's ability to bypass
national interests or to restructure EU interest representation can at
times be substantial, the analysis also found that the Commission's
attempts at framing the issues according to its objectives eventually
failed almost entirely. More generally, the empirical analysis calls the
analytical value of perceiving of EU institutions as single political
actors into question. Partially as a corollary to this point, the evidence
presented here conflicts with theoretical approaches that exclusively
conceptualise policy frames as organisationally embedded. Clearly, the
institutional organisation of politics is one of the most central sources
of competing frames of reference in the policy process. But policy
frames need not originate in organisations, and the dynamics a

salient framing conflict generates are often transformative in ways that transcend or even recast the legislative and administrative organisation of a policy making system. This, finally, leads back to the overarching theme of policy making as a process of conflict and consolidation. Schattschneider's dictum that policy conflicts can transform politics by creating new alliances was amply supported. But the EU policy process also proved alarmingly inept in containing the politicisation of issues once they had attracted the attention of the main institutional players. The empirical analysis presented here clearly failed to produce any evidence that framing conflicts in the EU mark only short and disruptive episodes, followed by prolonged periods of merely incremental adjustment. Instead of reaching a new equilibrium, framing conflicts became more vicious, volatile and persistent once they had risen to the top of the EU agenda. Given the EU's multiple constituencies, its complex system of representation and the fluctuating nature of its political conflict space, expansion and displacement of a policy conflict may sometimes be the only available mechanisms to establish stable policy equilibria in the EU. Identifying the mechanisms of politicisation through which these conflicts evolve and affect the political construct of interests in the European Union may be the main theoretical contribution of framing research in EU political analysis.

Literature

Agence Europe (1953–today) *Europe Bulletin Quotidien.* Luxembourg: Agence internationale d'information pour la presse.

Allison, Graham T. (1969) Conceptual Models and the Cuban Missile Crisis. *American Political Science Review* 63(3): 689–718.

Aspinwall, Mark and Gerald Schneider (2000) Same Menu, Separate Tables: The Institutional Turn in Political Science and the Study of European Integration. *European Journal of Political Research* 38(1): 1–36.

Bachrach, Peter and Morton Baratz (1963) Decisions and Nondecisions: An Analytic Framework. *American Political Science Review* 57(3): 632–642.

—— (1962) The Two Faces of Power. *American Political Science Review* 56: 947–952.

Bartolini, Stefano (2006) Mass Politics in Brussels: How Benign Could It Be? *Zeitschrift für Staats- und Europawissenschaften* 4(1): 28–56.

Bauer, Michael W. (2002) Limitations to Agency Control in European Union Policy-Making: The Commission and the Poverty Program. *Journal of Common Market Studies* 40(3): 381–400.

Baumgartner, Frank R. (2007) EU Lobbying: A View from the US. *Journal of European Public Policy* 14(3): 482–488.

—— (2001) Political Agendas, in Nelson Polsby (ed.) *International Encyclopaedia of Social and Behavioral Sciences: Political Science.* New York: Elsevier Science.

Baumgartner, Frank R. and Bryan D. Jones (2002) Positive and Negative Feedback in Politics, in Frank R. Baumgartner and Bryan D. Jones (eds) *Policy Dynamics.* Chicago: University of Chicago Press, pp. 3–28.

—— (1993) *Agendas and Instability in American Politics.* Chicago: University of Chicago Press.

—— (1991) Agenda Dynamics and Policy Subsystems. *The Journal of Politics* 53(4): 1044–1074.

Baumgartner, Frank R. and Christine Mahoney (2008) The Two Faces of Framing. Individual-Level Framing and Collective Issue Definition in the European Union. *European Union Politics* 9(3): 435–449.

Beyers, Jan and Bart Kerremans (2004) Bureaucrats, Politicians, and Societal Interests: How is European Policy Making Politicized? *Contemporary Political Studies* 37(10): 1119–1150.

BioTechnology (1995) BioTechnology Newsletter. Issue February 1995. Brussels: BIODOC Centre, DG Research.

Birkland, Thomas A. (1998) Focusing Events, Mobilization, and Agenda Setting. *Journal of Public Policy* 18(1): 53–74.

Bouwen, Pieter (2004) The Logic of Access to the European Parliament. *Journal of Common Market Studies* 42(3): 473–495.

Burns, Charlotte (2006) Co-decision and Inter-Committee Conflict in the European Parliament Post-Amsterdam. *Government and Opposition* 41(2): 230–248.

Cantley, Mark F. (1995) The Regulation of Modern Biotechnology: A Historical and European Perspective, in H.-J. Rehm and G. Reed (eds) *Biotechnology*. Weinheim: VCH Verlagsgesellschaft, pp. 505–631.

Christiansen, Thomas (1997) Tensions of European Governance: Politicised Bureaucracy and Multiple Accountability in the European Commission. *Journal of European Public Policy* 4(1): 73–90.

Christoforou, Theofanis (2004) The Regulation of Genetically Modified Organisms in the European Union: The Interplay of Science, Law and Politics. *Common Market Law Review* 41(3): 637–709.

Cini, Michelle (2006) The European Commission: An Unelected Legislator? *Journal of Legislative Studies* 8(4): 14–26.

Cobb, Roger W. and Charles D. Elder (1971) The Politics of Agenda-Building: An Alternative Perspective for Modern Democratic Theory. *Journal of Politics* 33(4): 892–915.

Cobb, Roger W., Jennie-Keith Ross and Marc H. Ross (1976) Agenda Building as a Comparative Political Process. *American Political Science Review* 70(1): 126–138.

Cobb, Roger W. and Marc H. Ross (eds) (1997) *Cultural Strategies of Agenda Denial*. Lawrence: University Press of Kansas.

Coen, David (1998) The European Business Interests and the Nation States: Large-Firm Lobbying in the European Union and Member States. *Journal of Public Policy* 18(1): 75–100.

Council of the European Union (1999) Draft Minutes of the 2194th Council Meeting (Environment) held in Luxembourg on 24 and 25 June 1999. 9433/1/99 REV. Brussels, 26 October 1999.

—— (1998) Council Regulation (EC) No. 1139/98 of 26 May 1998 Concerning the Compulsory Indication of the Labelling of Certain Foodstuffs Produced from Genetically Modified Organisms of Particulars Others Than Those Provided for in Directive 79/112/EEC. Brussels, 26 May 1998.

—— (1990a) Council Directive of 23 April 1990 on the Contained Use of Genetically Modified Micro-Organisms. 90/219/EEC. Brussels, 23 April 1990.

—— (1990b) Council Directive of 23 April 1990 on the Deliberate Release of Genetically Modified Organisms. 90/220/EEC. Brussels, 23 April 1990.

Cram, Laura (1998) The EU Institutions and Collective Action: Constructing a European Interest?, in Justin Greenwood and Mark Aspinwall (eds) *Collective Action in the European Union: Interests and the New Politics of Associability*. London: Routledge, pp. 63–80.

—— (1994) The European Commission as a Multi-Organisation: Social Policy and It Policy in the EU. *Journal of European Public Policy* 1(2): 195–217.

Crombez, Christophe (1997) The Co-Decision Procedure in the European Union. *Legislative Studies Quarterly* 22(1): 97–119.

—— (1996) Legislative Procedures in the European Community. *British Journal of Political Science* 26(2): 199–228.

Dery, David (2000) Agenda Setting and Problem Definition. *Policy Studies* 21(1): 37–47.

Dogan, Rhys (1997) Comitology: Little Procedures with Big Implications. *West European Politics* 20(3): 31–60.

Dudley, Geoffrey and Jeremy Richardson (1999) Competing Advocacy Coalitions and the Process of 'Frame Reflection': A Longitudinal Analysis of EU Steel Policy. *Journal of European Public Policy* 6(2): 225–248.

Earnshaw, David and Josephine Wood (1999) The European Parliament and Biotechnology Patenting: Harbinger of the Future? *Journal of Commercial Biotechnology* 5(4): 294–307.

Egeberg, Morton (2006) Executive Politics as Usual: Role Behaviour and Conflict Dimensions in the College of European Commissioners. *Journal of European Public Policy*, 13(1): 1–15.

—— (2004) An Organisational Approach to European Integration: Outline of a Complementary Perspective. *European Journal of Political Research* 43(2): 199–219.

—— (1996) Organization and Nationality in the European Commission Services. *Public Administration* 74(4): 721–735.

Eising, Reiner (2007) Institutional Context, Organizational Resources and Strategic Choices. Explaining Interest Group Access in the European Union. *European Union Politics* 8(3): 329–362.

—— (2004) Multilevel Governance and Business Interests in the European Union. *Governance* 17(2): 211–245.

Elster, Jon (1998) A Plea for Mechanisms, in Peter Hedström and Richard Swedberg (eds) *Social Mechanisms. An Analytical Approach to Social Theory.* Cambridge: Cambridge University Press, pp. 43–73.

Entman, Robert M. (1993) Framing: Toward Clarification of a Fractured Paradigm. *Journal of Communication* 43(4): 51–58.

EuropaBio (2001) EP Vote: Step Forward to Coherent Legal Framework for European Biotechnology. Brussels, 14 February 2001.

European Commission (2010a) Proposal for a Regulation of the European Parliament and Council Amending Directive 2001/18/EC as regards the possibility for the Member States to restrict or prohibit the cultivation of GMOs in their territory, COM(2010) 375 final. Brussels, 13 July 2010.

—— (2010b) Communication from the Commission to the European Parliament, the Council and the European Economic and Social Committee and the Committee of the Regions on the freedom for Member States to decide on the cultivation of genetically modified crops, COM(2010) 380 final. Brussels, 13 July 2010.

—— (2010c) Commission announces upcoming proposal on choice for Member States to cultivate or not GMO's and approves 5 decisions on GM's, IP10/222. Brussels, 2 March 2010.

—— (2010d) Commission Decision 2010/135/EU of 2 March 2010 concerning the placing on the market, in accordance with Directive 2001/18/EC of the European Parliament and of the Council, of a potato product (Solanum tuberosum L. line EH92-527-1) genetically modified for enhanced content of the amylopectin component of starch. Brussels, 2 March 2010.

—— (2009) Commission Decision 2009/244/EC of 16 March 2009 concerning the placing on the market, in accordance with Directive 2001/18/EC of the European Parliament and of the Council, of a carnation (Dianthus caryophyllus L., line 123.8.12) genetically modified for flower colour. Brussels, 16 March 2009.

—— (2007a) Commission Decision 2007/232/EC of 26 March 2007 concerning the placing on the market, in accordance with Directive 2001/18/EC of the European Parliament and of the Council, of oilseed rape products (Brassica napus L., lines Ms8, Rf3 and Ms8xRf3) genetically modified for tolerance to the herbicide glufosinate-ammonium. Brussels, 26 March 2007.

—— (2007b) Commission Decision 2007/364/EC of 23 May 2007 concerning the placing on the market, in accordance with Directive 2001/18/EC of the European Parliament and of the Council, of a carnation (Dianthus caryophyllus L., line 123.2.38) genetically modified for flower colour. Brussels, 23 May 2007.

—— (2007c) Second Report from the Commission to the Council and the European Parliament on the experience of Member States with GMOs placed on the market under Directive 2001/18/EC on the deliberate release into the environment of genetically modified organisms. COM(2007) 81 final. Brussels, 5 March 2007.

—— (2006) Commission Decision 2006/47/EC of 16 January 2006 concerning the placing on the market, in accordance with Directive 2001/18/EC of the European Parliament and of the Council, of a maize product (Zea mays L., hybrid MON 863 _ MON 810) genetically modified for resistance to corn rootworm and certain lepidopteran pests of maize. Brussels, 16 January 2006.

—— (2005a) Commission Decision 2005/772/EC of 3 November 2005 concerning the placing on the market, in accordance with Directive 2001/18/EC of the European Parliament and of the Council, of a maize product (Zea mays L., line 1507) genetically modified for resistance to certain lepidopteran pests and for tolerance to the herbicide glufosinate-ammonium. Brussels, 3 November 2005.

—— (2005b) Commission Decision 2005/635/EC of 31 August 2005 concerning the placing on the market, in accordance with Directive 2001/18/EC of the European Parliament and of the Council, of an oilseed rape product (Brassica napus L., GT73 line) genetically modified for tolerance to the herbicide glyphosate. Brussels, 31 August 2005.

—— (2005c) Commission Decision 2005/608/EC of 8 August 2005 concerning the placing on the market, in accordance with Directive 2001/18/EC of the European Parliament and of the Council, of a maize product (Zea mays L., line MON 863) genetically modified for resistance to corn rootworm. Brussels, 8 August 2005.

—— (2005d) GMOs: Commission reaction on Council votes on safeguards and GM maize MON863, IP/05/793. Brussels/Luxembourg, 24 June 2005.

—— (2004a) Commission Decision 2004/643/EC of 19 July 2004 concerning the placing on the market, in accordance with Directive 2001/18/EC of the European Parliament and of the Council, of a maize product (Zea mays L. line NK603) genetically modified for glyphosate tolerance. Brussels, 19 July 2004.

—— (2004b) Commission Regulation (EC) No 641/2004 of 6 April 2004 on detailed rules for the implementation of Regulation (EC) No 1829/2003 of the European Parliament and of the Council as regards the application for the authorisation of new genetically modified food and feed, the notification of existing products and adventitious or technically unavoidable presence of genetically modified material which has benefited from a favourable risk evaluation. Brussels, 6 April 2004.

—— (2004c) Commission Decision No 2004/204 of 23 February 2004 laying down detailed arrangements for the operation of the registers for recording information on genetic modifications in GMOs, provided for in Directive 2001/18/EC of the European Parliament and of the Council. Brussels, 23 February 2004.

—— (2004d) GMOs: Commission takes stock of progress, IP 04/118. Brussels. 28 January 2004.

—— (2004e) Commission Regulation (EC) No 65/2004 of 14 January 2004 establishing a system for the development and assignment of unique identifiers for genetically modified organisms. Brussels, 14 January 2004.

—— (2003a) European Legislative Framework for GMOs is Now in Place. IP/03/1056. Brussels, 22 July 2003.

—— (2003b) Internal Co-Ordination of Life Sciences and Biotechnology. Communication from the President to the Commission. Brussels, July 2003.

—— (2003c) Europeans and Biotechnology in 2002: A Report to the EC Directorate General for Research from the Project 'Life Sciences in European Society' (Eurobarometer 58.0) by George Gaskell, Nick Allum and Sally Stares. Brussels, March 2003.

—— (2002) Amended Proposal for a Regulation of the European Parliament and the Council on Genetically Modified Food and Feed. COM(2002) 559 final. Brussels, 8 October 2002.

—— (2001a) Proposal for a Regulation of the European Parliament and the Council on Genetically Modified Food and Feed. COM(2001) 425 final. Brussels, 25 July 2001.

—— (2001b) Proposal for a Regulation of the European Parliament and the Council Concerning Traceability and Labelling of Genetically Modified Organisms and Traceability of Food and Feed Products Produced from Genetically Modified Organisms and Amending Directive 2001/18/EC. COM(2001) 182 final. Brussels, 25 July 2001.

—— (2000a) Opinion of the Commission on the European Parliament's Amendments and the Council's Common Position Regarding the Proposal for a Directive Amending Directive 90/220/EEC on the Deliberate Release of Genetically Modified Organisms and Repealing Council Directive 90/220/EEC. COM(2000) 293 final. Brussels, 16 May 2000.

—— (2000b) White Paper on Food Safety. COM(1999) 719 final. Brussels, 12 January 2000.

—— (1999a) GMOs. IP/99/512. Brussels, 15 July 1999.

—— (1999b) Amended Proposal for a European Parliament and Council Directive Amending Directive 90/220/EEC on the Deliberate Release of Genetically Modified Organisms. COM(1999) 139 final. Brussels, 25 March 1999.

—— (1998a) Proposal for a European Parliament and Council Directive Amending Directive 90/220/EEC on the Deliberate Release of Genetically Modified Organisms. COM(1998) 85 final. Brussels, 23 February 1998.

—— (1998b) Proposal for a Council Regulation (EC) Concerning the Compulsory Indication on the Labelling of Certain Foodstuffs Produced from Genetically Modified Organisms of Particulars Other Than Those Provided for in Directive 79/112/EEC. COM(1998) 99 final. Brussels, 25 February 1998.

—— (1997a) The Commission Proposes Modification of Directive on Deliberate Release into the Environment of Genetically Modified Organisms. IP/97/1044. Brussels, 26 November 1997.

—— (1997b) The European Commission Approves the Labelling of Genetically Modified Organisms. IP/97/528 REV. Brussels, 18 June 1997.

—— (1997c) Commission Directive 97/35/EC of 18 June 1997 Adapting to Technical Progress for the Second Time Council Directive 90/220/EEC on the Deliberate Release into the Environment of Genetically Modified Organisms. Brussels, 18 June 1997.

—— (1997d) Reorganization of the Commission's Departments Responsible for Food Health. IP/97/112. Brussels, 12 February 1997.

—— (1997e) Consumer Health and Food Safety. Communication from the Commission. COM(1997) 183 final. Brussels, 30 April 1997.

—— (1996a) Report on the Review of the Directive 90/220/EEC in the Context of the Commission's Communication on Biotechnology and the White Paper. COM(96) 630 final. Brussels, 10 December 1996.

—— (1996b) Commission presents report on directive 90/220/EEC on genetically modified organisms. IP/96/1148. Brussels, 10 December 1996.

—— (1995) Proposal for a Council Directive Amending Directive 90/219/EEC on the Contained Use of Genetically Modified Micro-Organisms. COM(95) 640 final. Brussels, 6 December 1995.

—— (1994–2009) Bulletin of the European Union. Luxembourg: Office for Official Publications of the EC.

—— (1994) Biotechnology and the White Paper on Growth, Competitiveness and Employment. Preparing the Next Stage. COM(94) 219 final. Brussels, 1 July 1994.

—— (1993a) White Paper on Growth, Competitiveness and Employment. The Challenges and Ways Forward into the 21st Century. COM(93) 700 final. Brussels, 5 December 1993.

—— (1993b) Summary Record of a Round Table on the Biotechnology Regulatory Framework. Internal Coordination – Secretariat-General. Brussels, 4 October 1993.

—— (1992) Proposal for a Council Regulation (EEC) on Novel Foods and Food Ingredients. COM(92) 295 final. Brussels, 7 July 1992.

—— (1991) Promoting the Competitive Environment for the Industrial Activities Based on Biotechnology within the Community. SEC(91) 629 final. Brussels, 15 April 1991.

—— (1988a) Proposal for a Council Directive on the Contained Use of Genetically Modified Micro-Organisms. COM(88) 160-1 final. Brussels, 4 May 1988.

—— (1988b) Proposal for a Council Directive on the Deliberate Release of Genetically Modified Organisms. COM(88) 160-2 final. Brussels, 4 May 1988.

—— (1988c) Proposal for a Council Directive on the Legal Protection of Biotechnological Inventions. COM(88) 0496. Brussels, 17 October 1988.

—— (1986) A Community Framework for the Regulation of Biotechnology. COM(86) 573 final. Brussels, 4 November 1986.

—— (1985a) Speech by Mr Karl-Heinz Narjes at the Conference 'Industrial Biotechnology in Europe: Issues for Public Policy'. Brussels, 7 November 1985.

—— (1985b) The Commission's Approach to the Regulation of Biotechnology. Biotechnology Regulations Interservice Committee. Draft. Brussels, 28 October 1985.

—— (1983a) Biotechnology in the Community. COM(83) 672 final. Brussels, 3 October 1983.

—— (1983b) Biotechnology: The Community's Role. COM(83) 328 final. Brussels, 8 June 1983.

European Committee on Regulatory Aspects of Biotechnology (ECRAP) (1986) Safety and Regulation. European Committee on Regulatory Aspects of Biotechnology. Brussels, April 1986.

European Information Service (1972–today) Europe Environment. Brussels: European Information Service.

European Parliament (1999) Report on the Proposal for a European Parliament and Council Directive Amending Directive 90/220/EEC on the Deliberate Release into the Environment of Genetically Modified Organisms. Committee on the Environment, Public Health and Consumer Protection. PE 227.836/fin. Brussels, 28 January 1999.

—— (1996a) Report on the Commission Communication on Biotechnology and the White Paper on Growth, Competitiveness and Employment Preparing the Next Stage. Technological Development and Energy Committee on Research. PE 213.201 final. Brussels, 9 February 1996.

—— (1996b) Report on the Report from the Commission on the Review of Directive 90/220/EEC in the Context of the Commission's Communication on Biotechnology and the White Paper. Committee on the Environment, Public Health and Consumer Protection. PE 222.540/fin. Brussels, 3 July 1997.

—— (1994) Draft Report on Biotechnology within the Community. Committee on Energy, Research and Technology. PE 208.602/A. Brussels, 7 March 1994.

—— (1986) Resolution on Biotechnology in Europe and the Need for an Integrated Policy. Doc. A2–134/86. Brussels, 1986.

—— (1985) Committee on Energy, Research and Technology. Biotechnology Hearing: Outline. PE 98.227/rev. Brussels, 30 October 1985.

European Parliament and the Council of the European Union (2008) Directive 2008/27/EC of the European Parliament and of the Council of 11 March 2008 amending Directive 2001/18/EC on the deliberate release into the environment of genetically modified organisms, as regards the implementing powers conferred on the Commission. Brussels, 11 March 2008.

—— (2003a) Regulation (EC) No 1829/2003 of the European Parliament and of the Council of 22 September 2003 on Genetically Modified Food and Feed. Brussels, 22 September 2003.

—— (2003b) Regulation (EC) No 1830/2003 of the European Parliament and of the Council of 22 September 2003 Concerning the Traceability and

Labelling of Genetically Modified Organisms and the Traceability of Food and Feed Products Produced from Genetically Modified Organisms and Amending Directive 2001/18/EC. Brussels, 22 September 2003.

—— (2003c) Regulation (EC) No 1946/2003 of the European Parliament and of the Council of 15 July 2003 on Transboundary Movements of Genetically Modified Organisms. Brussels, 15 July 2003.

—— (2002) Regulation (EC) No 178/2002 of the European Parliament and the Council of 28 January 2002 Laying Down the General Principles and Requirements of Food Law, Establishing the European Food Safety Authority and Laying Down Procedures in Matters of Food Safety. Brussels, 28 January 2002.

—— (2001) Directive 2001/18/EC of the European Parliament and of the Council of 12 March 2001 on the Deliberate Release of Genetically Modified Organisms and Repealing Council Directive 90/220/EEC. Brussels, 12 March 2001.

—— (1998) Directive 98/44/EC of the European Parliament and of the Council of 6 July 1998 on the Legal Protection of Biotechnological Inventions. 98/44/EC. Brussels, 6 July 1998.

—— (1997) Regulation (EC) No 258/97 of the European Parliament and of the Council of 27 January 1997 Concerning Novel Foods and Food Ingredients. Brussels, 27 January 1997.

European Voice (1995–today) European Voice. Brussels: The Economist Group.

Francescon, Silvia (2001) The New Directive 2001/18/EC on the Deliberate Release of Genetically Modified Organisms into the Environment: Changes and Perspectives. *Review of European Community and International Environmental Law* 10(3): 309–320.

Gamson, William A. (1989) News as Framing. *American Behavioral Scientist* 33(2): 157–161.

George, Alexander L. and Andrew Bennett (2005) *Case Studies and Theory Development in the Social Sciences*. Cambridge: MIT Press.

Gerring, John (2005) Causation: A Unified Framework for the Social Sciences. *Journal of Theoretical Politics* 17(2): 163–198.

Greenwood, Justin and Karsten Ronit (1995) European Bioindustry, in Justin Greenwood (ed.) *European Casebook on Business Alliances*. London: Prentice Hall, pp. 75–85.

—— (1994) Interest Groups in the European Community: Newly Emerging Dynamics and Forms. *West European Politics* 17(1): 31–52.

—— (1992) Established and Emergent Sectors: Organised Interests at the European Level in the Pharmaceutical Industry and the New Biotechnologies, in Justin Greenwood, Jürgen R. Grote and Karsten Ronit (eds) *Organized Interests and the European Community*. London: Sage, pp. 69–98.

Harcourt, Alison J. (1998) EU Media Ownership Regulations: Conflict over Definition of Alternatives. *Journal of Common Market Studies* 36(3): 369–389.

Heclo, H. Hugh (1972) Review Article: Policy Analysis. *British Journal of Political Science* 2(1): 83–108.

Hedström, Peter and Richard Swedberg (1996) Social Mechanisms. *Acta Sociologica* 39(3): 281–308.

Hix, Simon (1998) The Study of the European Union: The 'New Governance' Agenda and Its Rivals. *Journal of European Public Policy* 5(1): 38–65.

Hix, Simon, Abdul Noury and Gérard Roland (2006) Dimensions of Politics in the European Parliament. *American Journal of Political Science* 50(2): 494–511.

Hörl, Björn, Andreas Warntjen and Arndt Wonka (2005) Build on Quicksand? A Decade of Procedural Spatial Models on EU Legislative Decision-Making. *Journal of European Public Policy* 12(3): 592–606.

Jones, Bryan D. (2001) *Politics and the Architecture of Choice. Bounded Rationality and Governance.* Chicago: University of Chicago Press.

—— (1999) Bounded Rationality. *Annual Review of Political Science* 2: 297–321.

—— (1994a) A Change of Mind or a Change of Focus? A Theory of Choice Reversals in Politics. *Journal of Public Administration Research and Theory* 4(2): 141–177.

—— (1994b) *Reconceiving Decision-Making in Democratic Politics. Attention, Choice, and Public Policy.* Chicago: University of Chicago Press.

Jones, Bryan D. and Frank R. Baumgartner (2002) Punctuations, Ideas, and Public Policy, in Frank R. Baumgartner and Bryan D. Jones (eds) *Policy Dynamics.* Chicago: University of Chicago Press, pp. 293–306.

Jupille, Joseph (2007) Contested Procedures: Ambiguities, Interstices and EU Institutional Change. *West European Politics* 30(2): 301–320.

Kahneman, Daniel (1997) New Challenges to the Rationality Assumption. *Legal Theory* 3(2): 105–124.

Kassim, Hussein (1994) Policy Networks, Networks, and European Union Policy-Making: A Sceptical View. *West European Politics* 17(4): 15–27.

Kellow, Aynsley (1989) Taking the Long Way Home? A Reply to Spitzer on the Arenas of Power. *Policy Studies Journal* 17(3): 537–546.

—— (1988) Promoting Elegance in Policy Theory: Simplifying Lowi's Arenas of Power. *Policy Studies Journal* 16(4): 713–724.

King, David C. (1994) The Nature of Congressional Committee Jurisdictions. *American Political Science Review* 88(1): 48–62.

Kreppel, Amie (2002) Moving Beyond Procedure: An Empirical Analysis of European Parliament Legislative Influence. *Comparative Political Studies* 35(7): 784–813.

Lawrence, Daniel, James Kennedy and Elizabeth Hattan (2002) New Controls on the Deliberate Release of Gmos. *European Environmental Law Review* 11(2): 51–56.

Lenschow, Andrea and Tony Zito (1998) Blurring or Shifting Policy Frames? Institutionalization of the Economic-Environmental Policy Linkage in the European Community. *Governance* 11(4): 415–441.

Lowi, Theodore J. (1972) Four Systems of Policy, Politics, and Choice. *Public Administration Review* 32(4): 298–310.

—— (1970) Decision Making vs. Policy Making: Toward and Antitode for Technocracy. *Public Administration Review* 30(3): 314–325.

—— (1964) American Business, Public Policy, Case-Studies, and Political Theory. *World Politics* 16(4): 677–715.

MacKenzie, Ruth and Silvia Francescon (2000) The Regulation of Genetically Modified Foods in the European Union: An Overview. *NYU Environmental Law Journal* 8(3): 530–555.

Mahoney, Christine (2004) The Power of Institutions: State and Interest Group Activity in the European Union. *European Union Politics* 5(4): 441–466.

Mahoney, Christine and Frank R. Baumgartner (2008) Converging Perspectives on Interest Group Research in Europe and America. *West European Politics* 31(6): 1253–1273.

Mahoney, James (2001) Beyond Correlational Analysis: Recent Innovations in Theory and Method. *Sociological Forum* 16(3): 575–593.

—— (2000) Strategies of Causal Inference in Small-N Analysis. *Sociological Methods and Research* 28(4): 387–424.

—— (1999) Nominal, Ordinal, and Narrative Appraisal in Macro-Causal Analysis. *American Journal of Sociology* 104(4): 1154–1196.

Mather, Janet (2001) The European Parliament: A Model of Representative Democracy? *West European Politics* 24(1): 181–201.

Mair, Peter (1997) E. E. Schattschneider's the Semisovereign People. *Political Studies* 45(5): 947–954.

Majone, Giandomenico (2003) The Politics of Regulation and European Regulatory Institutions, in Jack Hayward and Anand Menon (eds) *Governing Europe*. Oxford: Oxford University Press, pp. 297–312.

—— (2002) The European Commission: The Limits of Centralization and the Perils of Parliamentarization. *Governance* 15(3): 375–392.

—— (1994) The Rise of the Regulatory State in Europe. *West European Politics* 17(3): 78–102.

March, James G. (1994) *A Primer on Decision Making: How Decisions Happen*. New York: Free Press.

—— (1978) Bounded Rationality, Ambiguity, and the Engineering of Choice. *Bell Journal of Economics* 9(2): 587–608.

March, James G. and Herbert A. Simon (1958) *Organizations*. New York: Wiley.

Marks, Gary, Liesbet Hooghe and Kermit Blank (1996) European Integration from the 1980s: State-Centric Vs. Multi-Level Governance. *Journal of Common Market Studies* 34(3): 341–378.

Marks, Gary and Doug McAdam (1999) On the Relationship of Political Opportunities to the Form of Collective Action: The Case of the European Union, in Donatella della Porta, Hanspeter Kriesi and Dieter Rucht (eds) *Social Movements in a Globalizing World*. New York: St. Martin's Press, pp. 97–111.

Marks, Gary and Marco Steenbergen (2002) Understanding Political Contestation in the European Union. *Comparative Political Studies* 35(8): 879–892.

Mayntz, Renate (2004) Mechanisms in the Analysis of Social Macro-Phenomena. *Philosophy of the Social Sciences* 34(2): 237–259.

Moe, Terry M. (2005) Power and Political Institutions. *Perspectives on Politics* 3(2): 215–233.

Mörth, Ulrika (2000) Competing Frames in the European Commission – The Case of the Defence Industry and Equipment Issue. *Journal of European Public Policy* 7(2): 173–189.

Nelson, Barbara J. (1984) *Making an Issue of Child Abuse: Political Agenda Setting for Social Problems*. Chicago: University of Chicago Press.

Neuhold, Christine (2001) The 'Legislative Backbone' Keeping the Institution Upright? The Role of European Parliament Committees in the EU Policy-Making Process. *European Integration online Papers (EIoP)*: 5(10).

Nugent, Neill (1999) *The Government and Politics of the European Union*. Basingstoke: Macmillan.

—— (1998) Decisionmaking Procedures, in Desmond Dinan (ed.) *Encyclopedia of the European Union*. Boulder: Lynne Rienner, pp. 117–121.

—— (1997) At the Heart of the Union, in Neill Nugent (ed.) *At the Heart of the Union: Studies of the European Commission*. London: Macmillan Press, pp. 1–26.

—— (1995) The Leadership Capacity of the European Commission. *Journal of European Public Policy* 2(4): 603–623.

Nylander, Johan (2001) The Construction of a Market. A Frame Analysis of the Liberalization of the Electricity Market in the European Union. *European Societies* 3(3): 289–314.

Patterson, Lee Ann (1996) Biotechnology Policy: Regulating Risk and Risking Regulation, in Helen Wallace and William Wallace (eds) *Policy-Making in the European Union*. Oxford: Oxford University Press, pp. 317–343.

Peters, B. Guy (2005) The Problem of Policy Problems. *Journal of Comparative Policy Analysis* 7(4): 349–370.

—— (2001) Agenda Setting in the European Union, in Jeremy Richardson (ed.) *European Union. Power and Policymaking*. London: Routledge, pp. 77–94.

—— (1994) Agenda-Setting in the European Community. *Journal of European Public Policy* 1(1): 9–26.

—— (1992) Bureaucratic Politics and the Institutions of the European Community, in Alberta M. Sbragia (ed.) *Euro-Politics: Institutions and Policymaking in the 'New' European Community*. Washington: Brookings Institution, pp. 75–122.

Peterson, John (2004) Policy Networks, in A. Wiener and T. Diez (eds) *European Integration Theory*. Oxford: Oxford University Press, pp. 117–135.

Peterson, John (2001) The Choice for EU Theorists: Establishing a Common Framework for Analysis. *European Journal of Political Research* 39: 289–318.

—— (1995) Decision-Making in the European Union: Towards a Framework for Analysis. *Journal of European Public Policy* 2(1): 69–93.

Petracca, Mark P. (1992) Issue Definitions, Agenda-Building, and Policymaking. *Policy Currents* 2(3): 1–4.

Pierson, Paul (1993) When Effect Becomes Cause: Policy Feedback and Political Change. *World Politics* 45(4): 595–628.

Pijnenburg, Bert (1998) EU Lobbying by ad hoc Coalitions: An Exploratory Case Study. *Journal of European Public Policy* 5(2): 303–321.

Pollack, Mark A. (2005) Theorizing the European Union: International Organization, Domestic Polity, or Experiment in New Governance. *Annual Review of Political Science* 8: 357–398.

—— (2001) International Relations Theory and European Integration. *Journal of Common Market Studies* 39(2): 221–244.

—— (2000) The End of Creeping Competence? EU Policy-Making since Maastricht. *Journal of Common Market Studies* 38(3): 519–538.

—— (1998) The Engines of Integration? Supranational Authority and Influence in the European Union, in Wayne Sandholtz and Alec Stone Sweet (eds) *European Integration and Supranational Governance*. Oxford: Oxford University Press, pp. 217–249.

—— (1997a) Delegation, Agency and Agenda Setting in the EU. *International Organization* 51(1): 99–134.

—— (1997b) Representing Diffuse Interests in EC Policy-Making. *Journal of European Public Policy* 4(4): 572–590.

—— (1996) The New Institutionalism and EC Governance: The Promise and Limits of Institutional Analysis. *Governance* 9(4): 429–458.

Princen, Sebastiaan (2009) *Agenda-Setting in the European Union*. Houndmills, Basingstoke: Palgrave.

—— (2007) Agenda Setting in the European Union. A Theoretical Exploration and Agenda for Research. *Journal of European Public Policy* 14(1): 21–38.

Princen, Sebastiaan and Bart Kerremans (2008) Opportunity Structures in the EU Multi-Level System. *West European Politics* 31(6): 1129–1146.

Quattrone, George A. and Amos Tversky (1988) Contrasting Rational and Psychological Analysis of Political Choice. *American Journal of Political Science* 82(3): 719–736.

Radaelli, Claudio M. (1995a) The Role of Knowledge in the Policy Process. *Journal of European Public Policy* 2(2): 159–183.

—— (1995b) Corporate Direct Taxation in the European Union: Explaining the Policy Process. *Journal of Public Policy* 15(3 or 2): 153–181 or 43–71.

Raunio, Tapio (2000) Second-Rate Parties: Towards a Better Understanding of Europeans Parliament's Party Groups, in Knut Heidar and Rood Koole (eds) *Parliamentary Party Groups in European Democracies: Political Parties Behind Closed Doors*. London: Routledge, pp. 231–246.

Rein, Martin and Donald A. Schön (1997) Problem Setting in Policy Research, in Carol H. Weiss (ed.) *Using Social Research in Public Policy Making*. Lexington, MA: Lexington Books, pp. 235–251.

—— (1996) Frame-Critical Policy Analysis and Frame-Reflective Policy Practice. *Knowledge and Policy* 9(1): 85–104.

—— (1993) Reframing Policy Discourse, in Frank Fischer and John Forester (eds) *The Argumentative Turn in Policy Analysis and Planning*. Durham: Duke University Press, pp. 145–166.

—— (1991) Frame-Reflective Policy Discourse, in Peter Wagner, Carol H. Weiss, B. Wittrock and H. Wollman (eds) *Social Science, Modern States: National Experiences and Theoretical Crossroads*. Cambridge: Cambridge University Press, pp. 262–289.

Richardson, Jeremy (2000) Government, Interests Groups and Policy Change. *Political Studies* 48(5): 1006–1025.

Riker, William H. (1995) The Political Psychology of Rational Choice Theory. *Political Psychology* 16(1): 23–44.

—— (1990) Heresthetic and Rhetoric in the Spatial Model, in James M. Enelow and Melvin J. Hinich (eds) *Advances in the Spatial Theory of Voting*. Cambridge: Cambridge University Press, pp. 46–65.

—— (1986) *The Art of Political Manipulation*. New Haven: Yale University Press.
—— (1984) The Heresthetics of Constitution-Making: The Presidency in 1787, with Comments on Determinism and Rational Choice. *American Political Science Review* 78(1): 1–16.
—— (1982) The Two-Party System and Duverger's Law: An Essay on the History of Political Science. *American Political Science Review* 76(4): 753–766.
Ringe, Nils (2005) Policy Preference Formation in Legislative Politics: Structures, Actors, and Focal Points. *American Journal of Political Science* 49(4): 731–745.
Rittberger, Berthold (2000) Impatient Legislator and New Issue-Dimensions: A Critique of the Garrett-Tsebelis 'Standard Version' of Legislative Politics. *Journal of European Public Policy* 7(4): 554–575.
Rochefort, David A. and Roger W. Cobb (1994) Problem Definition: An Emerging Perspective, in David A. Rochefort and Roger W. Cobb (eds) *The Politics of Problem Definition: Shaping the Policy Agenda*. Lawrence: University Press of Kansas, pp. 1–31.
—— (1993) Problem Definition, Agenda Access, and Policy Choice. *Policy Studies Journal* 21(1): 56–71.
Sabatier, Paul A. (1998) The Advocacy Coalition Framework: Revisions and Relevance for Europe. *Journal of European Public Policy* 5(1): 98–130.
Scharpf, Fritz W. (2000) Institutions in Comparative Policy Research. *Comparative Political Studies* 33(6/7): 762–790.
Schattschneider, Elmer E. (1960) *The Semisovereign People: A Realist's View of Democracy in America*. Englewood Cliffs: Prentice-Hall.
—— (1957) Intensity, Visibility, Direction and Scope. *American Political Science Review* 51(4): 933–942.
—— (1935) *Politics, Pressures and the Tariff*. New York: Prentice-Hall.
Schendelen, Rinus van (2003) The GMO Food Arena in the EU (1998–2001). *Journal of Public Affairs* 3(3): 225–231.
Schmidt, Susanne K. (2000) Only an Agenda Setter? The European Commission's Power over the Council of Ministers. *European Union Politics* 1(1): 37–61.
Schmitter, Philippe C. (1992) Representation in the Future Euro-Polity. *Staatswissenschaften und Staatspraxis* 3(3): 379–405.
Schön, Donald A. and Martin Rein (1994) *Frame Reflection: Toward the Resolution of Intractable Policy Controversies*. New York: Basic Books.
Scully, Roger (2003) MEPs as Representatives: Individual and Institutional Roles. *Journal of Common Market Studies* 41(2): 269–288.
Sebenius, James K. (1983) Negotiation Arithmetic: Adding and Subtracting Issues and Parties. *International Organization* 37(2): 281–316.
Selck, Torsten J. (2004a) The European Parliament's Legislative Powers Reconsidered: Assessing the Current State of the Procedural Models Literature. *Politics* 24(2): 79–87.
—— (2004b) On the Dimensionality of European Union Legislative Decision-Making. *Journal of Theoretical Politics* 16(2): 203–222.
Senior Advisory Group on Biotechnology (SAGB) (1990a) *Community Policy for Biotechnology: Priorities and Actions*. Brussels: SAGB.
—— (1990b) *Community Policy for Biotechnology: Economic Benefits and European Competitiveness*. Brussels: SAGB.

188 *Literature*

Shepsle, Kenneth A. (1979) Institutional Arrangements and Equilibrium in Multi-Dimensional Voting Models. *American Journal of Political Science* 23(1): 27–59.

Sheridan, Brian (2001) *EU Biotechnology Law and Practice. Regulating Genetically Modified and Novel Food Products.* Isle of Wight: Palladian Law Publishing.

Simon, Herbert A. (1995) Rationality in Political Behavior. *Political Psychology* 16(1): 45–61.

—— (1987) Politics as Information Processing. *LSE Quarterly* 1(4): 345–370.

—— (1986) Rationality in Psychology and Economics. *Journal of Business* 59(4, Part 2): S209–S224.

—— (1985) Human Nature in Politics: The Dialogue of Psychology with Political Science. *American Political Science Review* 79(2): 293–304.

—— (1983) *Reason in Human Affairs.* Oxford: Basil Blackwell.

—— (1973) Applying Information Technology to Organization Design. *Public Administration Review* 33(3): 268–278.

Smyrl, Marc E. (1998) When (and How) Do the Commission's Preferences Matter? *Journal of Common Market Studies* 36(1): 79–99.

Spitzer, Robert J. (1989) From Complexity to Simplicity: More on Policy Theory and the Arenas of Power. *Policy Studies Journal* 17(3): 529–536.

—— (1987) Promoting Policy Theory: Revising the Arenas of Power. *Policy Studies Journal* 15(4): 675–689.

Steunenberg, Bernard (2000) Seeing What You Want to See: The Limits of Current Modelling on the European Union. *European Union Politics* 1(3): 368–373.

—— (1994) Decision-Making under Different Institutional Arrangements: Legislation by the European Community. *Journal of Theoretical and Institutional Economics* 150(4): 642–669.

Stone, Deborah A. (1989) Causal Stories and the Formation of Policy Agendas. *Political Science Quarterly* 104(2): 281–300.

Tallberg, Jonas (2000) The Anatomy of Autonomy: An Institutional Account of Variation in Supranational Influence. *Journal of Common Market Studies* 38(5): 843–864.

Thomson, Robert, Jovanka Boerefijn and Frans Stokman (2004) Actor Alignments in European Union Decision-Making. *European Journal of Political Research* 43(2): 237–261.

Toke, Dave (2004) *The Politics of GM Food.* London: Routledge.

Traxler, Franz and Philippe C. Schmitter (1995) The Emerging Euro-Polity and Organized Interests. *European Journal of International Relations* 1(2): 191–218.

Trondal, Jarle (2008) Balancing Roles of Representation in the European Commission. *Acta Politica* 43(4): 429–452.

—— (2007a) Contending Decision-Making Dynamics within the European Commission. *Comparative European Politics* 5(2): 158–178.

—— (2007b) The Public Administration Turn in Integration Research. *Journal of European Public Policy* 14(6): 960–972.

Tsebelis, George (1994) The Power of the European Parliament as a Conditional Agenda Setter. *American Political Science Review* 88(1): 128–142.

Tsebelis, George and Geoffrey Garrett (2001) The Institutional Foundations of Intergovernmentalism and Supranationalism in the European Union. *International Organization* 55(2): 357–390.

—— (2000) Legislative Politics in the European Union. *European Union Politics* 1(1): 9–36.

Tsioumani, Elsa (2004) Genetically Modified Organisms in the EU: Public Attitudes and Regulatory Developments. *Review of European Community and International Environmental Law* 13(3): 279–288.

Tversky, Amos and Daniel Kahneman (1986) Rational Choice and the Framing of Decisions. *The Journal of Business* 59(4, Part 2): S251–S278.

—— (1981) The Framing of Decisions and the Psychology of Choice. *Science* 211: 453–458.

Weiss, Janet A. (1989) The Powers of Problem Definition: The Case of Government Paperwork. *Policy Sciences* 22(2): 97–121.

Wendon, Bryan (1998) The Commission as Image-Venue Entrepreneur in EU Social Policy. *Journal of European Public Policy* 5(2): 339–353.

Whitaker, Richard (2005) National Parties in the European Parliament: An Influence in the Committee System? *European Union Politics* 6(5): 5–28.

Wilson, James Q. (1980) The Politics of Regulation, in James Q. Wilson (ed.) *The Politics of Regulation*. New York: Basic Books, pp. 357–394.

—— (1973) Organizations and Public Policy, in James Q. Wilson (ed.) *Political Organizations*. New York: Basic Books, pp. 327–347.

Woll, Cornelia (2006) Lobbying in the European Union: From Sui Generis to a Comparative Perspective. *Journal of European Public Policy* 13(3): 456–469.

Wonka, Arndt (2008) Decision-Making Dynamics in the European Commission: Partisan, National or Sectoral? *Journal of European Public Policy* 15(8): 1145–1163.

Zimmer, Christina, Gerald Schneider and Michael Dobbins (2005) The Contested Council: Conflict Dimensions of an Intergovernmental EU Institution. *Political Studies* 53(2): 403–422.

Index